**TECHNICAL ENGINEERING AND DESIGN GUIDES AS ADAPTED FROM THE U.S. ARMY CORPS OF ENGINEERS, NO. 29**

# Topographic Surveying

Published by
**ASCE PRESS**

American Society of Civil Engineers
1801 Alexander Bell Drive
Reston, VA 20191-4400

Abstract: This manual establishes procedural guidance, specifications, and quality control criteria for performing field topographic surveying in support of planning, engineering and design, construction, and environmental restoration activities. It covers field survey techniques used in performing topographic surveys with modern electronic total stations and electronic data collectors. It includes procedures for transferring field data to computer-aided drafting and design (CADD) systems or geographic information systems (GIS).

Library of Congress Cataloging-in-Publication Data

Topographic surveying.
    p.   cm. — (Technical engineering and design guides as adapted from the U.S. Army Corps of Engineers ; no. 29)
    Includes bibliographical references and index.
    ISBN 0-7844-0374-0
    1. Topographical surveying—Handbooks, manuals, etc. I. American Society of Civil Engineers. II. Series.
TA590.T699        1999
526.9'8—dc21                                                                                                                                                                          99-049722

The material presented in this publication has been prepared in accordance with generally recognized engineering principles and practices, and is for general information only. This information should not be used without first securing competent advice with respect to its suitability for any general or specific application.

The contents of this publication are not intended to be and should not be construed to be a standard of the American Society of Civil Engineers (ASCE) and are not intended for use as a reference in purchase specifications, contracts, regulations, statutes, or any other legal document.

No reference made in this publication to any specific method, product, process or service constitutes or implies an endorsement, recommendation, or warranty thereof by ASCE.

ASCE makes no representation or warranty of any kind, whether express or implied, concerning the accuracy, completeness, suitability, or utility of any information, apparatus, product, or process discussed in this publication, and assumes no liability therefore.

Anyone utilizing this information assumes all liability arising from such use, including but not limited to infringement of any patent or patents.

Photocopies. Authorization to photocopy material for internal use or personal use under circumstances not falling within the fair use provisions of the Copyright Act is granted by ASCE to libraries and other users registered with the Copyright Clearance Center (CCC) Transactional Reporting Service, provided that the base fee of $8.00 per article plus $.50 per page copied is paid directly to CCC, 222 Rosewood Drive, Danvers, MA 01923. The identification for ASCE Books is 0-7844-0374-0/00/$8.00 = $.50. Requests for special permission or bulk copying should be addressed to Permissions & Copyright Dept., ASCE.

Copyright © 2000 by the American Society of Civil Engineers,
exclusive of U.S. Army Corps of Engineers material.
All Rights Reserved.
Library of Congress Catalog Card No: 99-049722
ISBN 0-7844-0374-0
Manufactured in the United States of America.

# TABLE OF CONTENTS

## Chapter 1. Introduction

   1-1. Purpose ........................................................ 1
   1-2. Applicability .................................................. 1
   1-3. References .................................................... 1
   1-4. Scope of Manual ............................................... 1
   1-5. Metrics ....................................................... 1
   1-6. Brand Names ................................................... 1
   1-7. Accompanying Guide Specifications .............................. 2
   1-8. USACE Capabilities ............................................ 2
   1-9. COGO System ................................................... 2
   1-10. Sample Scope of Work ......................................... 2
   1-11. Glossary ..................................................... 2
   1-12. Manual Development and Proponency ............................ 2

## Chapter 2. Topographic Accuracy Standards

   2-1. General ....................................................... 3
   2-2. Topographic Mapping Standards ................................. 3
   2-3. USACE Topographic Mapping Standard ............................ 8
   2-4. Intended Use of the Map ....................................... 8
   2-5. Area of the Project ........................................... 8
   2-6. Map Scale ..................................................... 8
   2-7. Contour Interval .............................................. 9
   2-8. ASPRS Accuracy Standards ...................................... 9
   2-9. USACE Horizontal Accuracy Check .............................. 11

## Chapter 3. Topographic Survey Control

   3-1. General ...................................................... 12
   3-2. USACE Control Survey Accuracy Standards ...................... 12
   3-3. Reconnaissance and Planning Phase ............................ 13
   3-4. Primary Survey Control ....................................... 14
   3-5. GPS Survey Control ........................................... 14
   3-6. Secondary Control for Topographic Surveys .................... 14
   3-7. Plane Coordinate Systems ..................................... 14
   3-8. Scale Factor Considerations .................................. 15
   3-9. Control Checks ............................................... 16

## Chapter 4. Topographic Survey Techniques

   4-1. General ...................................................... 20
   4-2. Engineering Site Plan Surveys ................................ 20

4-3. Utility Surveys ... 21
4-4. As-Built Surveys ... 21

**Section I. Plane-Table Surveys**
    4-5. General ... 22
    4-6. Plane-Table Topography ... 22
    4-7. Plane-Table Triangulation ... 22
    4-8. Plane-Table Resection ... 23
    4-9. Plane-Table Two-Point Problem ... 24
    4-10. Plane-Table Traverse ... 25
    4-11. Plane-Table Stadia Traverse ... 25
    4-12. Plane-Table Three-Point Orientation ... 25
    4-13. Contouring Methods ... 26
    4-14. Locating and Plotting Detail ... 27
    4-15. Plane-Table Equipment Checklist ... 28
    4-16. Plane-Table Setup Hints ... 28
    4-17. Plane-Table Notekeeping ... 29
    4-18. Plane-Table Location Details ... 29

**Section II. Electronic Total Station Surveys**
    4-19. Electronic Total Stations ... 30
    4-20. Field Equipment ... 31
    4-21. Equipment Maintenance ... 32
    4-22. Maintaining Battery Power ... 32
    4-23. Total Station Job Planning and Estimating ... 33
    4-24. Electronic Theodolite Error Sources ... 34
    4-25. Total Survey System Error Sources and How To Avoid Them ... 35
    4-26. Controlling Errors ... 36
    4-27. Coding Field Data ... 37
    4-28. Field Computers ... 38
    4-29. Modem for Data Transfer (Field to Office) ... 38
    4-30. Trigonometric Leveling and Vertical Traversing ... 38
    4-31. Trigonometric Leveling Field Procedures ... 39
    4-32. Trigonometric Leveling Error Sources ... 39

# Chapter 5. Data Collection Procedures for the Total Station

    5-1. General ... 41
    5-2. Functional Requirements of a Generic Data Collector ... 41
    5-3. Data Collection Operating Procedures ... 42
    5-4. Field Crew Responsibility ... 46
    5-5. Surveyor–Data Collector Interface ... 48
    5-6. Digital Data ... 48
    5-7. Digital Transfer ... 48
    5-8. Data Collector Requirements ... 50
    5-9. Coding Field Data ... 50
    5-10. Summary of Total Station FIELD-TO-FINISH Procedures ... 51
    5-11. Data Collectors ... 51

# Chapter 6. Surveyor Data Collector Interface and Formats

    6-1. Computer Interfacing ... 54
    6-2. Data Standardization ... 54
    6-3. Coordinate File Coding ... 54

|       |                                                    |    |
|-------|----------------------------------------------------|----|
| 6-4.  | Data Sets                                          | 56 |
| 6-5.  | Computer-Aided Design and Drafting (CADD) Interface| 57 |
| 6-6.  | Total Station Data Collection and Input to CADD    | 57 |
| 6-7.  | CADD Plotting                                      | 57 |

## Chapter 7. Map Compilation

| | | |
|---|---|---|
| 7-1. | General | 58 |
| 7-2. | CVTPC | 58 |
| 7-3. | File Descriptions | 59 |
| 7-4. | Overview of Topographic Survey Data Flow | 59 |
| 7-5. | Typical Point Descriptors Used in Topographic Surveying | 59 |

## Chapter 8. Architect-Engineer Contracts

| | | |
|---|---|---|
| 8-1. | General | 65 |
| 8-2. | Preparation | 65 |
| 8-3. | Scope of Work | 65 |

## Chapter 9. Route Surveying

| | | |
|---|---|---|
| 9-1. | General | 66 |
| 9-2. | Horizontal Circular Curves | 66 |
| 9-3. | Deflection Angles | 66 |
| 9-4. | Degree of Curve—Arc Definition | 67 |
| 9-5. | Degree of Curve—Chord Definition | 67 |
| 9-6. | Curve Stakeouts | 67 |
| 9-7. | Curve Formulas | 67 |
| 9-8. | Transition Spirals | 68 |
| 9-9. | Spiral Stakeout | 68 |
| 9-10. | Vertical Curves | 69 |
| 9-11. | Vertical Curve—Tangent Offset Method | 69 |
| 9-12. | Vertical Curve—Equation Method | 70 |
| 9-13. | Vertical Curve Obstructions | 70 |

## Appendix A. References

| | | |
|---|---|---|
| A-1. | Required Publications | 71 |

## Appendix B. Guide Specification for Topographic Mapping Services

**Instructions**

| | | |
|---|---|---|
| B-1. | General | 72 |
| B-2. | Coverage | 72 |
| B-3. | Applicability | 72 |
| B-4. | Contract Format | 72 |
| B-5. | General Guide Use | 72 |
| B-6. | Insertion of Technical Specifications | 73 |
| B-7. | Alternate Clauses/Provisions or Options | 73 |
| B-8. | Notes and Comments | 73 |
| B-9. | IDT Contracts and Individual Work Order Assignments | 73 |

**Appendix C. Automated Topographic Survey Data Collector Equipment: Inventory of USACE Instrumentation and Software** .......................... 86

**Appendix D. Coordinate Geometry Software**

    D-1. General ............................................................. 88
    D-2. Requirements ...................................................... 88
    D-3. Functions .......................................................... 88

**Appendix E. Sample Scope of Work** ........................................ 90

**Appendix F. Glossary** ..................................................... 94

**Index** .................................................................... 95

**DEPARTMENT OF THE ARMY**
U.S. Army Corps of Engineers
WASHINGTON, D.C. 20314-1000

September 28, 1999

REPLY TO
ATTENTION OF:

**Engineering and Construction Division**

Mr. Delon Hampton
President, American Society
 of Civil Engineers
345 East 47th Street
New York, New York 10017

Dear Mr. Hampton:

   I am pleased to furnish the American Society of Civil Engineers (ASCE) a copy of the U. S. Army Corps of Engineers Engineering Manual, EM 1110-1-1005, Topographic Surveying. The Corps uses this manual to provide procedural guidance, technical specifications and quality control criteria for topographic surveying.

   I understand that ASCE plans to publish this manual for public distribution. I believe this will benefit the civil engineering community by improving transfer of technology between the Corps and other engineering professionals.

                                        Sincerely,

                                        Joe N. Ballard
                                        Lieutenant General, USA
                                        Commanding

## American Society of Civil Engineers

Technical Engineering and Design Guides as adapted from the US Army Corps of Engineers

- No. 1: Design of Pile Foundations
- No. 2: Strength Design for Reinforced-Concrete Hydraulic Structures
- No. 3: Design, Construction, and Maintenance of Relief Wells
- No. 4: Retaining and Flood Walls
- No. 5: Roller-Compacted Concrete
- No. 6: Coastal Groins and Nearshore Breakwaters
- No. 7: Bearing Capacity of Soils
- No. 8: Standard Practice for Concrete for Civil Works Structures
- No. 9: Settlement Analysis
- No. 10: Hydraulic Design of Flood Control Channels
- No. 11: Standard Practice for Shotcrete
- No. 12: Hydraulic Design of Spillways
- No. 13: Construction with Large Stone
- No. 14: Photogrammetric Mapping
- No. 15: Design of Sheet Pile Walls
- No. 16: Rock Foundations
- No. 17: Monitor Well Design, Installation, and Documentation of Hazardous and/or Toxic Waste Sites
- No. 18: River Hydraulics
- No. 19: Flood-Runoff Analysis
- No. 20: Channel Stability Assessment for Flood Control Projects
- No. 21: Structural Design of Closure Structures for Local Flood Protection Projects
- No. 22: Design of Hydraulic Steel Structures
- No. 23: Geophysical Exploration for Engineering and Environmental Investigations
- No. 24: Chemical Grouting
- No. 25: Hydrographic Surveying
- No. 26: Instrumentation of Embankment Dams and Levees
- No. 27: Construction Control for Earth and Rockfill Dams
- No. 28: NAVSTAR Global Positioning System Surveying
- No. 29: Topographic Surveying

# CHAPTER 1

# INTRODUCTION

## 1-1. Purpose

This manual establishes procedural guidance, specifications, and quality control criteria for performing field topographic surveying in support of planning, engineering and design, construction, and environmental restoration activities.

## 1-2. Applicability

This manual applies to all HQUSACE elements, major subordinate commands, and districts that perform, contract, or monitor topographic surveys in support of civil works and military construction activities. It is also applicable to surveys performed or procured by local interest groups under various cooperative or cost-sharing agreements.

## 1-3. References

Required and related references are listed in Appendix A.

## 1-4. Scope of Manual

This manual establishes standard procedures, minimum accuracy requirements, instrumentation and equipment requirements, and quality control criteria for performing field topographic surveys. It shall be used as a guide in planning and performing topographic surveys with USACE hired-labor forces. The manual has been written to include the electronic surveying methods that have changed conventional equipment and procedures in topographic surveying. Traditional methods, such as plane-table surveying, are included because these methods remain effective topographic surveying methods. Accuracy specifications, procedural criteria, and quality control requirements contained in this manual should be directly referenced in the scopes of work for Architect-Engineer (A-E) survey services or other third-party survey services to ensure that uniform and standardized procedures are followed by both hired labor and contract service sources throughout USACE.

Throughout the manual, topographic survey criteria standards are in specific terms and are normally summarized in tables. Guidance is in more general terms, where methodologies are described in readily available references or survey instrumentation operating manuals. Where procedural guidance is otherwise unavailable, it is provided herein. Sample computations and survey recordation formats are shown for some of the more common field operations.

The manual primarily focuses on the preparation of design drawings and other documents associated with engineering projects, including related contracted construction performance activities. Topographic mapping using photogrammetry or remote sensing methods is not covered in this manual (see EM 1110-1-1000).

## 1-5. Metrics

The use of both the metric and English systems of measurement in this manual is predicated due to the common use of both systems throughout the surveying and mapping profession. Spatial location coordinates are almost universally expressed in feet. Construction measurement quantities are normally measured in linear feet, square feet, or cubic yards. Spatial coordinates relative to the North American Datum of 1983 (NAD 83) are usually represented in metric units (International System of Units [SI]). Universal Transverse Mercator (UTM) projection coordinates are usually metric as well. Because of the variety of mixed measurements, equivalent conversions are not shown in this manual—the most common measurement unit is used for example computations. Most metric conversions are based exclusively on the U.S. Survey Foot, which equals (exactly) 1,200/3,937 m (or 3.280833333333 ft/m). The SI conversion (1 International Foot = exactly 30.48/100 m) is used in a few states.

## 1-6. Brand Names

The citation in this report of brand names of commercially available products does not constitute official endorsement or approval of the use of such products.

## 1-7. Accompanying Guide Specifications

This manual is designed to be used in conjunction with guide specifications in Appendix B as a quality control and quality assurance aid in administering A-E contracts for topographic surveying services.

## 1-8. USACE Capabilities

An inventory of USACE instrumentation and hardware is given in Appendix C.

## 1-9. COGO System

The coordinate geometry (COGO) software system is described in Appendix D.

## 1-10. Sample Scope of Work

A sample scope of work for an A-E contract is shown in Appendix E.

## 1-11. Glossary

Abbreviations used in this manual are explained in Appendix F.

## 1-12. Manual Development and Proponency

The HQUSACE proponent for this manual is the Surveying and Analysis Section, General Engineering Branch, Civil Works Directorate. Recommended corrections or modifications to this manual should be directed to:

HQUSACE
ATTN: CECW-EP-S
20 Massachusetts Ave., NW
Washington, DC 20314-1000

# CHAPTER 2

# TOPOGRAPHIC ACCURACY STANDARDS

## 2-1. General

This chapter sets forth the accuracy standards to be used in USACE for topographic mapping. The mapping accuracy standards are associated with the scales and sheet size of the finished map. Horizontal accuracy is directly related to the map scale. Vertical accuracy is a stated fraction of the contour interval. The contour interval is related to the vertical scale. Details of these map accuracies are stated in this chapter. The map standards set forth in this chapter shall have precedence over numbers, figures, references, or guidance presented in other chapters. USACE topographic surveying and mapping criteria are detailed in Table 2-1. Upon selection of the type of project to be mapped, the criteria limits are specified. The specific map scale and contour interval within these limits are selected according to specific project parameters. Survey accuracies needed to achieve these map accuracies are separate issues and are addressed in Chapter 3.

### a. Mapping standards

A map accuracy is determined by comparing the mapped location of selected well-defined points to their "true" location as determined by a conventional field survey. A map accuracy standard classifies a map as statistically meeting a certain level of accuracy. Horizontal (or planimetric) map accuracy standards are usually expressed in terms of two-dimensional radial positional error measures—the root-mean-square (RMS) statistic. Vertical map accuracy standards are in terms of one-dimensional RMS elevation errors. Map accuracy classifications are dependent on the specified (i.e., designed) target scale and vertical relief, or contour interval, of the map. See EM 1110-1-1000 and the FGCS *Multipurpose Land Information System Guidebook* for more detailed information.

### b. Surveying Standards

All maps warranting an accuracy classification must be referenced to, or controlled by, conventional field surveys. The surveying standards are independent of these map accuracy standards; survey accuracies based on relative closure estimates cannot necessarily be correlated with map accuracy positional error estimates. Survey accuracy is a function of the specifications and procedures used, the resultant internal or external closures, and is independent of the map scale or map contour interval. The accuracy of the conventional field survey used to test the map accuracy must exceed that of the map.

### c. Target Scale and Contour Interval Specifications

Mapping accuracy standards are associated with the final development scale of the map, both the horizontal "target" scale and vertical relief (specified contour interval) components. Photogrammetric mapping flight altitudes or ground topographic (topo) survey accuracy and density requirements are specified based on the design map target scale and contour interval. The use of computer-aided drafting and design (CADD) or Geographic Information Systems (GIS) equipment allows planimetric features and topographic elevations to be readily separated onto various layers and depiction at any scale. Problems arise when target scales are increased beyond their original values, or when so-called "rubber sheeting" is performed. Therefore, it is critical that these spatial data layers contain descriptor information identifying the original source target scale and designed accuracy.

## 2-2. Topographic Mapping Standards

Six generally recognized industry standards can be used for specifying spatial mapping products and resultant accuracy compliance criteria:

**Table 2-1. USACE Surveying and Mapping Requirements for Military Construction, Civil Works, Operations, Maintenance, Real Estate, and HTRW Projects**

| Project or Activity | Typical Target (Plot) Map Scale[a] 1 in. = X (ft) | Feature Location Tolerance[b] (ft-RMS) | USACE Control Survey[c] Accuracy | Feature Elevation Tolerance[b,d] (ft-RMS) | USACE Control Survey[c] Accuracy | Typical Contour Interval (ft) | ASPRS Map Accuracy Class |
|---|---|---|---|---|---|---|---|
| **MILITARY CONSTRUCTION (MCA, MCAF, OMA, OMAF):** | | | | | | | |
| **Design and Construction of New Facilities:** Site Plan Data for Direct Input into CADD 2D/3D Design Files | | | | | | | |
| General Site Plan Feature and Topo Detail | 30–50 | 0.1–0.5 | 3rd—I | 0.1–0.3 | 3rd | 1 | 1 or 2 |
| Surface/Subsurface Utility Detail | 30–50 | 0.2–0.5 | 3rd—I | 0.1–0.2 | 3rd | NA | 1 |
| Building or Structure Design | 20–50 | 0.05–0.2 | 3rd—I | 0.1–0.3 | 3rd | 1 | 1 |
| Airfield Pavement Design Detail | 20–40 | 0.05–0.1 | 3rd—I | 0.05–0.1 | 2nd | 0.5–1 | 1 |
| Grading and Excavation Plans (roads, drainage, etc.) | 30–100 | 0.5–2 | 3rd—I or II | 0.2–1 | 3rd | 1–2 | 2 |
| **M&R or Renovation of Existing Structures, Roadways, Utilities, etc.:** for Design/Construction P&S | 30–50 | 0.1–0.5 | 3rd—I | 0.1–0.5 | 3rd | 1 | 1 |
| **Recreational Site P&S**: (Golf Courses, Athletic Fields, etc.) | 100 | 1–2 | 3rd—II | 0.2–2 | 3rd | 2–5 | 2 |
| **Training Sites, Ranges, Cantonment Areas, etc.** | 100–200 | 1–5 | 3rd—II | 1–5 | 3rd | 2 | 2 or 3 |
| **Installation Master Planning and Facilities Management Activities** (including AM/FM and GIS Feature Applications) | | | | | | | |
| General Location Maps; for master planning purposes | 100–400 | 2–10 | 3rd—II | 1–10 | 3rd | 2–10 | 2 or 3 |
| Space Management (interior design/layout) | 10–50 | 0.05–1 | Relative to Structure | NA | NA | NA | NA |
| Installation Surface/Subsurface Utility Maps (As-Builts: Fuel Gas, Electricity, Communications, Cable, Storm Water, Sanitary, Water Supply, Treatment Facilities, Meters, etc.) | 50–100 (DA) 50 (USAF) | 0.2–1 | 3rd—I | 0.1 | 3rd | 1 | 1 |
| **Architectural Drawings** (see ER 1110-345-710): | | | | | | | |
| Floor Plans, Roof Plans, Exterior Elevations, Cross Sections | (1/16–1/4 in. per ft) | — | — | — | — | — | — |
| Wall Sections | (1/2 in. per ft) | — | — | — | — | — | — |
| Stair Details | (3/4 in. per ft) | — | — | — | — | — | — |
| Detail Plans | (3/4–3 in. per ft) | — | — | — | — | — | — |
| **Area/Installation/Base-Wide Mapping Control Network to Support Overall GIS and AM/FM Development**[e] | NA | varies 1–200 | 3rd—I or 2nd—II | varies 1–10 | 2nd or 3rd | 1–10 | 2 or 3 |
| Housing Management (Family Housing, Schools, Boundaries, and Other Installation Community Services) | 100–400 | 10–50 | 4th | NA | 4th | NA | 3 |
| Environmental Mapping and Assessments | 200–400 | 10–50 | 4th | NA | 4th | NA | 3 |

*continued on next page*

# TOPOGRAPHIC ACCURACY STANDARDS

**Table 2-1. (Continued) USACE Surveying and Mapping Requirements for Military Construction, Civil Works, Operations, Maintenance, Real Estate, and HTRW Projects**

| Project or Activity | Typical Target (Plot) Map Scale[a] 1 in. = X (ft) | Feature Location Tolerance[b] (ft-RMS) | USACE Control Survey[c] Accuracy | Feature Elevation Tolerance[b,d] (ft-RMS) | USACE Control Survey[c] Accuracy | Typical Contour Interval (ft) | ASPRS Map Accuracy Class |
|---|---|---|---|---|---|---|---|
| Emergency Services (Military Police, Crime/Accident Locations, Emergency Transport Routes, Post Security Zoning, etc.) | 400–2,000 | 50–100 | 4th | NA | 4th | NA | 3 |
| Cultural, Social, Historical (Other Natural Resources) | 400 | 20–100 | 4th | NA | 4th | NA | 3 |
| Runway Approach and Transition Zones; General Plans/Sections[f] | 100–200 | 5–10 | 3rd–II | 2–5 | 3rd | 5 | 2 or 3 |
| **CIVIL WORKS DESIGN, CONSTRUCTION, OPERATIONS, AND MAINTENANCE ACTIVITIES:** | | | | | | | |
| **Site Plan Mapping for Design Memoranda, Contract Plans and Specifications, etc.** (for Input to CADD 2D/3D Design Files) | | | | | | | |
| Locks, Dams, Flood Control Structures; Detail Design Plans | 20–50 | 0.05–1 | 2nd–II | 0.01–0.5 | 2nd or 3rd | 0.5–1 | 1 |
| Grading/Excavation Plans | 100 | 0.5–2 | 3rd–I | 0.2–1 | 3rd | 1–5 | 1 |
| Spillways, Concrete Channels, Upland Disposal Areas | 50–100 | 0.1–2 | 2nd–II | 0.2–2 | 3rd | 1–5 | 1 |
| Construction In-Place Volume Measurement | 40–100 | 0.5–2 | 3rd–I | 0.5–1 | 3rd | NA | 1 |
| **River and Harbor Navigation Projects:** Site Plan Mapping Design, Operation, or Maintenance of Flood Control Structures, Canals, Channels, etc. — for Contract Plans or Reports | | | | | | | |
| Levees and Groins (New Work or Maintenance Design Drawings) | 100 | 1–2 | 3rd–II | 0.5–1 | 3rd | 1–2 | 2 |
| Canals and Waterway Dredging (New Work Base Mapping)[g] | 100 | 2 | 3rd–II | 0.5 | 3rd | 1 | 3 |
| Canals and Waterway Dredging (Maintenance Drawings) | 200 | 2 | 3rd–II | 0.5 | 3rd | 1 | 3 |
| Beach Renourishment/Hurricane Protection Projects | 100–200 | 2 | 3rd–II | 0.5–1 | 3rd | 1 | 2 |
| Project Condition Reports: (Base Mapping for Plotting Hydrographic Surveys—Line Maps or Airphoto Plans) | 200–1,000 | 5–50 | 3rd–II | 0.5–1 | 3rd | 1–2 | 3 |
| Revetment Clearing, Grading, and As-Built Protection | 100–400 | 2–10 | 3rd–II | 0.5–1 | 3rd | 1–2 | 3 |
| **Geotechnical and Hydrographic Site Investigation Surveying Accuracies:** | | | | | | | |
| Hydrographic Contract Payment and P&S Surveys | — | 10 | NA | 0.5 | NA | 1 | NA |
| Hydrographic Project Condition Surveys | — | 20 | NA | 1.0 | NA | 1 | NA |
| Hydrographic Reconnaissance Surveys | — | 300 | NA | 1.5 | NA | 1 | NA |
| Geotechnical Investigative Core Borings | — | 5–10 | 4th | 0.1–0.5 | 3rd or 4th | 1–5 | NA |

*continued on next page*

Table 2-1. (Continued) USACE Surveying and Mapping Requirements for Military Construction, Civil Works, Operations, Maintenance, Real Estate, and HTRW Projects

| Project or Activity | Typical Target (Plot) Map Scale[a] 1 in. = X (ft) | Feature Location Tolerance[b] (ft-RMS) | USACE Control Survey[c] Accuracy | Feature Elevation Tolerance[b,d] (ft-RMS) | USACE Control Survey[c] Accuracy | Typical Contour Interval (ft) | ASPRS Map Accuracy Class |
|---|---|---|---|---|---|---|---|
| **General Planning and Feasibility Studies, Reconnaissance Reports, Permit Applications, etc.** | 100–400 | 2–10 | 3rd—II | 0.5–2 | 3rd | 2–10 | 3 |
| **Civil Works Projects—GIS Feature Mapping:** | | | | | | | |
| Area/Project-Wide Mapping Control Network to Support Overall GIS Development | NA | Varies 1–100 | 2nd—I or 2nd—II | Varies 1–10 | 2nd | 1–10 | 2 or 3 |
| Soil and Geologic Classification Maps, Well Points | 400 | 20–100 | 4th | NA | 4th | NA | 3 |
| Cultural and Economic Resources, Historic Preservation | 1,000 | 50–100 | 4th | NA | 4th | NA | 3 |
| Land Utilization GIS Classifications; Regulatory Permit General Locations | 400–1,000 | 50–100 | 4th | NA | 4th | NA | 3 |
| Socioeconomic GIS Classifications | 1,000 | 100 | 4th | NA | 4th | NA | 3 |
| Land Cover Classification Maps | 400–1,000 | 50–200 | 4th | NA | 4th | NA | 3 |
| **Archeological or Structure Site Detail Mapping** (including nontopographic—close range—photogrammetric mapping) | 0.5–10 | 0.01–0.5 | 2nd—I or II | 0.01–0.5 | 2nd | 0.1–1 | 1 |
| **Structural Deformation Monitoring Studies/Surveys[h]:** | | | | | | | |
| Reinforced Concrete Structures (Locks, Dams, Gates, Intake Structures, Tunnels, Penstocks, Spillways, Bridges, etc.) | Large-scale vector movement diagrams or tabulations | 0.001 | NA[i] | 0.001 | NA[i] | (Long-Term Movement) 0.01–0.1 | NA |
| Earth/Rockfill Structures (Dams, Floodwalls, Levees, etc.) | | 0.01–0.2 | NA | 0.01–0.1 | NA | 0.1 | NA |
| **Flood Control and Multipurpose Project Planning, Floodplain Mapping, Water Quality Analysis, and Flood Control Studies** | 400–1,000 | 20–100 | 3rd—I | 0.2–2 | 2nd or 3rd | 2–5 | 3 |
| **FEMA Flood Insurance Studies** | 400 | 20 | 3rd—I | 0.5 | 3rd | 4 | 3 |
| **REAL ESTATE ACTIVITIES (ACQUISITION, DISPOSAL, MANAGEMENT, AUDIT)[j]** | | | | | | | |
| **Tract Maps, Individual:** Detailing Installation or Reservation Boundaries, Lots, Parcels, Adjoining Parcels and Record Plats, Utilities, etc. | 20–400[k] | 0.05–2 | 3rd—I or II | 0.1–2 | 3rd | 1–5 | 1 |
| **Condemnation Exhibit Maps** | 20–400 | 0.05–2 | 3rd—I or II | 0.1–2 | 3rd | 1–5 | 1 |
| **Guide Taking Lines** (for Fee and Easement Acquisition) **Boundary Encroachment Maps** | 20–100 | 0.1–1 | 3rd—I or II | 0.1–1 | 3rd | 1 | 1 |
| **Real Estate GIS or LIS General Feature Mapping:** Land Utilization and Management, Forestry Management, Mineral Acquisition | 200–1,000 | 50–100 | 4th | NA | 4th | NA | 3 |
| **General Location or Planning Maps** | 1:24,000 (USGS) | 50–100 | NA | 5–10 | 3rd | 5–10 | – |
| **Easement Areas and Easement Delineation Lines** | 100 | 0.1–0.5 | 3rd—I or II | 0.1–0.5 | 3rd | – | 2 |

*continued on next page*

# TOPOGRAPHIC ACCURACY STANDARDS

**Table 2-1. (Continued) USACE Surveying and Mapping Requirements for Military Construction, Civil Works, Operations, Maintenance, Real Estate, and HTRW Projects**

| Project or Activity | Typical Target (Plot) Map Scale[a] 1 in. = X (ft) | Feature Location Tolerance[b] (ft-RMS) | USACE Control Survey[c] Accuracy | Feature Elevation Tolerance[b,d] (ft-RMS) | USACE Control Survey[c] Accuracy | Typical Contour Interval (ft) | ASPRS Map Accuracy Class |
|---|---|---|---|---|---|---|---|
| **HAZARDOUS AND TOXIC WASTE (HTW) SITE INVESTIGATION, MODELING, AND CLEAN–UP ACTIVITIES** | | | | | | | |
| **General Detailed Site Plan Mapping** (HTW Sites, Asbestos, etc.) | 5–50 | 0.2–1 | 2nd–II | 0.1–0.5 | 2nd or 3rd | 0.5–1 | 1 |
| **Subsurface Geotoxic Data Mapping** (Modeling) | 20–100 | 1–5 | 3rd–II | 1–2 | 3rd | 1–2 | 1 |
| **Contaminated Groundwater Plume Mapping** (Modeling) | 20–100 | 2–10 | 3rd–II | 1–5 | 3rd | 1–2 | 2 |
| **General HTW Site Planning, Reconnaissance Mapping, etc.** | 50–400 | 2–20 | 3rd–II | 2–20 | 3rd | 2–5 | 2 |
| **EMERGENCY OPERATION MANAGEMENT ACTIVITIES:** (Use basic GIS database requirements defined above.) | | | | | | | |

*Note:* ASPRS = American Society of Photogrammetry and Remote Sensing; OMA = Operations and Maintenance, Army; OMAF = Operations and Maintenance, Air Force; RMS = root-mean-square; HTRW = Hazardous, Toxic, and Radioactive Waste; M&R = Maintenance and Repair; P&S = Plans and Specifications; AM/FM = automated mapping/facilities management; GIS = Geographic Information System; LIS = Land Information System; NA = not applicable; CADD = computer-aided drafting and design; FEMA = Federal Emergency Management Administration.

[a] Target map scale is that contained in CADD, GIS, and/or AM/FM layer, and/or to which ground topo or aerial photography accuracy specifications are developed. This scale may not always be compatible with the feature location/elevation tolerances required—in many instances, design or real property features are located to a far greater relative accuracy than that which can be scaled at the target (plot) scale, such as property corners, utility alignments, first-floor or invert elevations, etc. Coordinates/elevations for such items are usually directly input into a CADD or AM/FM database.

[b] The map location tolerance (or precision) of a planimetric feature is defined relative to two adjacent points within the confines of a structure or map sheet, not to the overall project or installation boundaries. Relative accuracies are determined between two points which must functionally maintain a given accuracy tolerance between themselves, such as adjacent property corners, adjacent utility lines, adjoining buildings, bridge piers or approaches or abutments, overall building or structure site construction limits, runway ends, catch basins, levee baseline sections, etc. Tolerances between the two points are determined from the end functional requirements of the project/structure (e.g., field construction/fabrication, field stakeout or layout, alignment, locationing, etc.). Few engineering, construction, or real estate projects require that relative accuracies be rigidly maintained beyond a 5,000-ft range, and usually only within the range of the detailed design drawing for a project/structure (or its equivalent CADD design file limit). For example, two catch basins 200 ft apart should be located to 0.1 ft relative to each other but need only be known to ±100 ft relative to another catch basin 5 mi away. Likewise, relative accuracy tolerances are far less critical for small-scale GIS, LIS, and AM/FM data elements. Actual construction alignment and grade stakeout will generally be performed to the 0.1- or 0.01-ft levels, depending on the type of construction.

[c] USACE control survey accuracy refers to the procedural and closure specifications needed to obtain/maintain the relative accuracy tolerances needed between two functionally adjacent points on the map or structure, for construction or layout. Usually 3rd-order control procedures (horizontal and vertical) will provide sufficient accuracy for most work, and in many instances of small-scale mapping or GIS rasters, 3rd-order—Class II methods and 4th-order topo/construction control methods may be used. Base- or area-wide mapping control procedures shall be designed and specified to meet functional accuracy tolerances within the limits of the structure, building, or utility distance involved for design, construction, or real estate surveys. Higher-order control surveys shall not be specified for area-wide mapping or GIS definition unless a definitive functional requirement exists (e.g., military operational targeting or some low-gradient flood control projects).

[d] Some flood-control projects may require better relative accuracy tolerances than those shown.

[e] GIS raster or vector features generally can be scaled or digitized from any existing map of the installation—typically a standard USGS 1 in. = 2,000 ft scale quadrangle map is adequate given the low relative accuracies needed between GIS data features, elements, or classifications. Relative or absolute GPS positioning (10–300 ft) may be adequate to tie GIS features where no maps exist. In general, a basic area or installation-wide 2nd- or 3rd-order control network is adequate for all subsequent engineering, construction, real estate, GIS, and/or AM/FM control.

[f] Typical requirements for general approach maps are 1:50,000 (H) and 1:1,000 (V), detail maps at 1:5,000 (H) and 1:250 (V).

[g] Table refers to base maps upon which subsurface hydrographic surveys are plotted, not to hydrographic survey control.

[h] Long-term structural movements measured from points external to the structure may be tabulated or plotted in either X-Y-Z or by single vector movement normal to a potential failure plane. See EM 1110-2-4300 and EM 1110-2-1908 for stress–strain, pressure, seismic, and other precise structural deflection measurement methods within/between structural members, monoliths, cells, embankments, etc.

[i] Accuracy standards and procedures for structural deformation surveys are contained in EM 1110-1-1004. Horizontal and vertical deformation monitoring survey procedures are performed relative to a control network established for the structure. Ties to the National Geodetic Reference System (NGRS) or National Geodetic Vertical Datum of 1929 (NGVD 29) are not necessary other than for general reference and then need only USACE 3rd-order connection.

[j] Real property surveys shall conform to local/state minimum technical standards where prescribed by law or code.

[k] A scale of 1 in. = 100 ft is recommended by ER 405-1-12. Smaller scales should be on even 100-ft increments.

- Office of Management and Budget (OMB) United States National Map Accuracy Standards
- American Society of Photogrammetry (ASP) Specifications for Aerial Surveys and Mapping by Photogrammetric Methods for Highways
- U.S. Department of Transportation (DOT) Surveying and Mapping Manual Map Standards
- American Society of Photogrammetry and Remote Sensing (ASPRS) Accuracy Standards for Large-Scale Maps
- American Society of Civil Engineers (ASCE) Surveying and Mapping Division Standards
- U.S. National Cartographic Standards for Spatial Accuracy

Each of these standards has applications to different types of functional products, ranging from wide-area small-scale mapping (OMB National Map Accuracy Standards) to large-scale engineering design (ASPRS Accuracy Standards for Large-Scale Maps). Their resultant accuracy criteria (i.e., spatial errors in $x$, $y$, and $z$), including quality control compliance procedures, do not differ significantly from one another. In general, use of any of these standards will result in a quality map.

## 2-3. USACE Topographic Mapping Standard

The recommended standard for USACE topographic mapping is the ASPRS Accuracy Standards for Large-Scale Maps. This standard was developed (and is generally recognized) by the photogrammetric industry. The associated scale is defined for maps larger than 1:20,000 (1 in. = 1,667 ft). The scale range of most USACE large-scale topographic work is 1 in. = 10 ft to 1 in. = 200 ft. Maps for flood control and emergency services may have smaller scales. Topographic surveys in support of architectural drawing details may have larger scales. The ASPRS standards contain definitive statistical map testing criteria that can be used to truth a map. Tangible information for contract administration may be documented in a contract based on these testing criteria. For USACE small-scale maps, the OMB United States National Map Accuracy Standards are used. USACE map scales for these standards are ≤1:20,000. Maps generated at these scales will generally be flown by aerial photography. For details of this standard consult EM 1110-1-1000.

## 2-4. Intended Use of the Map

Table 2-1 lists recommended scales, contour intervals, and all associated position tolerances for USACE projects or activities. Functional activities are divided into military construction, civil works, real estate, hazardous waste, and emergency management. Sub-activities for each of these categories define the necessary map parameters. Use of Table 2-1 saves preliminary mapping research and establishes standards for USACE mapping requirements. Standards are especially important because of the high demand of digital data information. For most projects, identification of the type of project is the only design assumption required. USACE mapping parameters are selected across the appropriate row. The remaining sections of this chapter list criteria for narrowing a map design parameter for cases in Table 2-1 where a range is allowable. Map clarity, map cost, and map sheet size are considerations for narrowing parameter ranges to specific numbers in each category for a given project.

## 2-5. Area of the Project

The locations of points in a large area may be measured with consistent precision throughout, but the relative precision of the points located farthest from the control will tend to have more error than points located directly from control monuments. To maintain the required accuracy for a project, a primary project control net or loop is established to cover the entire project. Secondary project control loops or nets are constructed from the primary project network. This helps to ensure that the intended precision will not drop below the tolerance of the survey. In lieu of increasing control requirements, the map scale may be reduced. This trade-off between survey control and scale has either increased project costs or the scale has been reduced below usable limits in some cases. To resolve the trade-off problem, the ASPRS has stated the map accuracy relative to the finished map sheet. This substitutes relative survey line accuracies between points in the national network for relative survey line accuracies between points contained within the sheet borders. Map recipient requirements are met per sheet, which is usually the purpose of the majority of site plan mapping used in construction.

## 2-6. Map Scale

Map scale is the ratio of the distance measurement between two identifiable points on a map to the same physical points existing at ground scale. The errors in map plotting and scaling should exceed errors in measurements on the ground by a ratio of about 3 to 1. Stated in a different manner, a ratio can be established as a function of the plotter error divided by the allowable scale error. For example, if a digital plotter has an accuracy of ±0.25 mm and scaled map distances must be accurate to ±0.5 ft (152 mm), then 0.25/152 = 1/610; or the ratio becomes 1:600 or 1 in. = 50 ft.

Another common number used by surveyors to determine map scales and survey precision is an error of 1/40 in. (0.64 mm) between any two points scaled from the finished map. This error is assumed constant regardless of the length of a line until the scale is changed. For example, given a scale error of 1/40 in. and a feature accuracy requirement of ±10 ft, the maximum allowable map scale would be 1/40 in. per 10 ft, or a scale of 1 in. = 400 ft.

The traditional 1/40-in. plotting/scaling error probably originated from the National Map Accuracy Standard (NMAS). The NMAS specified not more than 10% of well-defined points (a group sample) tested in the field on a given map shall be in error by more than 1/30 in. (85 mm) for scales greater than 1:20,000 (large scale). Not more than 10% of points tested shall be in error exceeding 1/50 in. (50 mm) for scales ≤1:20,000. These measurements were tested at the publication scale for horizontal map truthing. Vertical map truthing specified not more than 10% of the elevations tested will exceed one-half the contour interval. The maximum error for a (one) plottable well-defined point, easily visible or recoverable on the ground, is 1/100 in. (0.25 mm) as defined by the NMAS. Note that the smaller the sample size, the more restrictive the tolerance becomes. This is why 20 points are required in the ASPRS standard. To compute the NMAS Circular Map Accuracy Standard (CMAS) from the ASPRS values, use the following conversion:

$$CMAS = 2.146 \times \sigma_{x \, or \, y}$$

The ASPRS standard emphasizes that the standard is based at full ground scale, and the CMAS can be compared with the 1/30-in. NMAS only approximately.

The surveyor should always use the smallest scale that will provide the necessary detail for a given project. This will provide economy and meet the project requirements. Use this rule-of-thumb when deciding limits as provided in Table 2-1. Once the smallest scale has been selected from Table 2-1, determine whether any other map uses that need a larger scale are possible for this project. If no other uses are of practical value, then the map scale has been determined.

## 2-7. Contour Interval

The contour interval is the constant elevation difference between two adjacent contour lines. The contour interval is chosen based on the map purpose, required vertical accuracy (if any was specified), the relief of the area of concern, and somewhat the map scale. Steep slopes (large relief) will cause the surveyor to increase the contour interval to make the map more legible. Flat areas will tend to decrease the interval to a limit that does not interfere with planimetric details located on the topographic map.

As a general rule, the lower limit for the contour interval is 25 lines per inch for even the smallest map scales. The checklist to find the proper contour interval follows:

1. intended purpose of the map
2. desired accuracy of the depicted vertical information
3. area relief (mountainous, hilly, rolling, flat, etc.)
4. cost of extra field work and possibility of plotting problems for selecting a smaller contour interval
5. other practical uses for the intended map

According to the above checklist, contour interval ranges are recommended in Table 2-1 for the kinds of projects typically encountered in USACE. If a specific vertical tolerance has been specified as the purpose for the mapping project, then the contour interval may be determined as a direct proportion from Table 2-1 for the type of project site. Otherwise, the stated map accuracy of the vertical information will be in terms of the selected contour interval within the limits listed in Table 2-1.

Any contour drawn on the map will be correct to a stated fraction of the selected contour interval. Because interpolation is used between spot elevations, the spot elevations themselves are required to be twice as precise as the contours generated by the spot elevations.

## 2-8. ASPRS Accuracy Standards

USACE has adopted the ASPRS accuracy specifications for large-scale mapping. The maps are divided into three classes. Class 1 holds the highest accuracies. Site plans for construction are this category. Class 2 has one-half the overall accuracy of Class 1. Typical projects include excavation, road grading, or disposal operations. Class 3 has one-third the accuracy (or three times the allowable error) of Class 1 maps. Large-area cadastral, city planning, and land classification maps are typically in this category. The ASPRS map class selection is listed for each activity or project type. Tables 2-2 and 2-3 detail ASPRS horizontal and vertical accuracy requirements, respectively.

A limiting RMS error (RMSE) for each class is indicated in Tables 2-2 and 2-3. The RMSE is found by locating prominent features by rectangular coordinates from a finished planimetric or topographic map. At least 20 check points are measured from the map. The points are selected in an agreement between the map producer and the client. A survey party then locates the same points on the ground. Third-order survey methods are sufficient in most cases, again depending on the map scale and the area of the project. The survey methods used for map testing must be superior to the methods used to construct the map in order to establish a truth basis.

To test horizontal features, planimetric coordinates of well-defined points are scaled from the finished map in ground scale units and subtracted from the same actual coordinates obtained during the field check survey. The test checks the x and y directions separately. The planimetric coordinate differences are inspected for any discrepancies that exceed three times the limiting RMSE according to class in Table 2-2. If more than 20 points were selected for the check survey, the discrepancies in excess of three times the RMS may be thrown out; but the entire point must be discarded (x,y,z). Outlier values existing in a minimum data set of 20 must be resolved in the field. Outlier values are considered blunders and shall be resolved before an ASPRS accuracy statement is printed on the finished map. The discrepancies are squared. Squares are summed and divided by the number of points used for the sample. The square root operation is performed on value obtained from the sum of the squared differences divided by the sample number to obtain the RMSE test statistic. This same procedure is then performed for the y and z coordinates.

Table 2-2 lists the planimetric feature coordinate accuracy requirements for well-defined points in either the x- or y-coordinate directions. The values in the columns under the classes are in ground units of feet. The values are at ground scale, not map scale. An accuracy statement in the notes of each map sheet shall be published stating that the finished map meets the USACE requirement for location of planimetric features at the published scale. The RMSE is computed for each map sheet produced and compared with Table 2-3 according to map class and target map scale.

The values of the RMSE differences are inversely proportional to the map scale. As the map scale decreases, the limiting RMSE in the differences increases. The increase in the allowable tolerance includes the total errors of survey control, data collection, map compilation/plotting, and scaling errors in depicting well-defined points from the finished map. The limiting accuracies listed in Table 2-2

**Table 2-2. Planimetric Feature Coordinate Accuracy Requirement (Ground X or Y) for Well-Defined Points**

| Target Map Scale | | Limiting RMS Error in X or Y (ft) ASPRS | | |
|---|---|---|---|---|
| 1 in. = x (ft) | Ratio (ft/ft) | Class 1 | Class 2 | Class 3 |
| 5 | 1:60 | 0.05 | 0.10 | 0.15 |
| 10 | 1:120 | 0.10 | 0.20 | 0.30 |
| 20 | 1:240 | 0.2 | 0.4 | 0.6 |
| 30 | 1:360 | 0.3 | 0.6 | 0.9 |
| 40 | 1:480 | 0.4 | 0.8 | 1.2 |
| 50 | 1:600 | 0.5 | 1.0 | 1.5 |
| 60 | 1:720 | 0.6 | 1.2 | 1.8 |
| 100 | 1:1,200 | 1.0 | 2.0 | 3.0 |
| 200 | 1:2,400 | 2.0 | 4.0 | 6.0 |
| 400 | 1:4,800 | 4.0 | 8.0 | 12.0 |
| 500 | 1:6,000 | 5.0 | 10.0 | 15.0 |
| 800 | 1:9,600 | 8.0 | 16.0 | 24.0 |
| 1,000 | 1:12,000 | 10.0 | 20.0 | 30.0 |
| 1,667 | 1:20,000 | 16.7 | 33.3 | 50.0 |

are the maximum permissible RMSE allowed under the ASPRS standard. Work performed by experienced topographic survey crews should be well within the limits of the values given in Table 2-2. As stated above, the RMSE is computed from a sample of no less than 20 well-defined points. Whether this acceptance testing is actually performed is a contracting officer determination—not all map products need to be tested. If the test is performed correctly and the map cannot meet the requirements of Table 2-2, the accuracy statement cannot be published in the notes until the map has been corrected by the map producer.

A vertical accuracy statement, based on the ASPRS map class and contour interval shown in Table 2-3, shall be included in the notes section of each map sheet produced for the USACE. The check used to meet the class requirement is based on the RMSE and tolerance limits of Table 2-3. Table 2-3 lists two kinds of values based on a given contour interval. The first group of values (left side of the table) correspond to topographic map features and the allowable vertical error in the finished map. These features may be roads, buildings, trees, hilltops, swales, valleys, or creeks. The other values (right side of the table) are allowable errors for spot locations, which are actual topographic locations where measurements were taken. Spot locations

# TOPOGRAPHIC ACCURACY STANDARDS

**Table 2-3. ASPRS Topographic Elevation Accuracy Requirement for Well-Defined Points**

| Target Contour Interval (ft) | ASPRS Limiting RMS Error (ft) | | | | | |
|---|---|---|---|---|---|---|
| | Topo Feature Points | | | Spot or DTM Elevation Points | | |
| | Class 1 | Class 2 | Class 3 | Class 1 | Class 2 | Class 3 |
| 0.5 | 0.17 | 0.33 | 0.50 | 0.08 | 0.16 | 0.25 |
| 1 | 0.33 | 0.66 | 1.00 | 0.17 | 0.33 | 0.50 |
| 2 | 0.67 | 1.33 | 2.00 | 0.33 | 0.67 | 1.00 |
| 4 | 1.33 | 2.67 | 4.00 | 0.67 | 1.33 | 2.00 |
| 5 | 1.67 | 3.33 | 5.00 | 0.83 | 1.67 | 2.50 |

*Note:* DTM = digital terrain model.

require one-half the allowable error of their corresponding feature points for a given contour interval.

## 2-9. USACE Horizontal Accuracy Check

In Table 2-1, note *b* explains horizontal map accuracy in terms of relative accuracies of structures critical to engineering or construction objectives. This is the intention of most mapping produced in USACE. For construction applications, a minimum of three pairs of points (total of six points) should be tested per map sheet. These points should be located in the construction area and be intended for length and orientation of significant existing structures. The planimetric features used for identifying pairs on a finished map will be well-defined man-made points.

Field locations of well-defined project points are recorded such that distances can be chained at the time of the survey. The procedures used are typical of a building location survey. Coordinates are computed for the locations and checked against the chained distances to ensure that blunders or systematic errors are minimized.

The same points are scaled or digitized in the office from the finished map sheet. For each of the six points, $\Delta x$ and $\Delta y$ are computed. Taking the six points two at a time generates a combination of 15 errors from the map sheet. Calculation of the 15 errors is easily done on a computer. The computation should be performed in terms of latitude and departure instead of distance alone. The sum of the squares of the latitude and departure errors will result in an RMSE for comparison with the tolerances shown in Table 2-1.

# CHAPTER 3

# TOPOGRAPHIC SURVEY CONTROL

## 3-1. General

USACE control work can generally be separated into primary project control and secondary project control. Primary control, usually on the State Plane Coordinate System (SPCS) projection, is intended to accommodate all project-related surveying tasks identified for the life of a particular project. Secondary control is intended for more specific applications or densification, such as topographic surveying or construction stakeout. The specified USACE control accuracies are listed by activity type in Table 2-1. To explain check surveys designed to establish map accuracies, see ASPRS Accuracy Specification for Large-Scale Line Maps (in EM 1110-1-1000).

USACE primary project control is sometimes tied into, but rarely adjusted as part of, the National Geodetic Reference System (NGRS). This tie is generally performed by converting geodetic control coordinates into the SPCS or Universal Transverse Mercator (UTM). The grid coordinates of the control monuments are inversed for grid azimuth and grid distance. The azimuth is used as a starting azimuth (or check azimuth) and the ground-scale distance is used to check the monuments.

The Global Positioning System (GPS) has significantly increased the production of surveys performed by surveyors. The USACE may occasionally require a geodetic survey because of unique construction requirements or project size, but most projects are surveyed based on SPCS or UTM horizontal grids. Independent survey datums or reference systems should not be used for primary project control unless required by local code, statute, or practice. These include local tangent planes, state High Accuracy Regional Networks (HARNs), and unreferenced construction baseline station-offset control. The horizontal datum for the primary control should be the North American Datum of 1983 (NAD 83/86). New projects should be based on the NAD 83 system and some older projects may remain based on the NAD 27 system.

## 3-2. USACE Control Survey Accuracy Standards

### a. Horizontal Control Standards

The USACE uses the relative point closure survey accuracy standard. The compass rule adjustment procedure will produce this result by dividing the closure by the sum of the traverse lengths. Table 3-1 lists the classification for closure standards.

**Horizontal Point Closure.** The horizontal point closure is determined by dividing the linear distance misclosure of the survey into the overall circuit length of a traverse, loop, or network line/circuit. When independent directions or angles are observed, as on a conventional survey (i.e., traverse, trilateration, or triangulation), these angular misclosures may optionally be distributed before assessing positional misclosure. The horizontal point closure is also used in USACE as an accuracy measure. In cases where GPS vectors are measured in geocentric coordinates, then the three-dimensional positional misclosure is assessed.

**Approximate Surveying.** Approximate surveying work should be classified based on the survey's estimated or observed positional errors. This would include absolute GPS and some differential GPS techniques with positional accuracies ranging from 10 to 150 ft (2 standard deviations root-mean-

Table 3-1. USACE Point Closure Standards for Horizontal Control Surveys

| USACE Classification | Point Closure Standard Ratio |
|---|---|
| Second-Order Class I | 1:50,000 |
| Second-Order Class II | 1:20,000 |
| Third-Order Class I | 1:10,000 |
| Third-Order Class II | 1:5,000 |
| Fourth-Order—Construction Layout | 1:2,500 to 1:20,000 |

square [2DRMS]). There is no order classification for such approximate work.

**Higher Order Surveys.** Requirements for relative accuracies or closures exceeding 1:50,000 are rare for most USACE applications. Recommended control survey accuracies based on the functional project application are listed in Table 2-1. Surveys requiring accuracies of first order (1:100,000) or better should be performed using Federal Geodetic Control Subcommittee (FGCS) standards and specifications.

### b. Vertical Control Standards

The vertical accuracy of a survey is determined by the elevation misclosure within a level section or level loop. For conventional differential or trigonometric leveling, section or loop misclosures (in feet) shall not exceed the limits listed in Table 3-2, where the line or circuit length ($M$) is measured in miles. Procedural specifications or restrictions pertaining to vertical control surveying methods or equipment should not be over-restrictive.

## 3-3. Reconnaissance and Planning Phase

The reconnaissance phase could be the most important phase of the survey. At this phase of the project, the required control accuracy is known. Topographic maps, aerial photographs, tax maps, and basically any mapping information are collected for the area. These maps, along with required site visits, shall be used to extract slopes, soil characteristics, ground cover, drainage structures, nearby utilities, and other physical evidence. Weather information and rainfall data should be consulted at this time. Additionally, wetlands, historical artifacts, and other like sites shall be identified. All the above information should be consolidated into a reconnaissance report. Photographs are encouraged in this report. Ownership information (name and address) and legal descriptions of privately owned lands should be obtained during this phase. Right-of-entry should be secured on lands not owned or controlled by the government. Permits may be required and should be obtained at this time.

Vertical information is important for horizontal control computations as well as the project vertical control, in general. U.S. Geological Survey (USGS) quadrangles and publications of the National Ocean Service (NOS) horizontal/vertical control are invaluable to the beginning of the project. Elevations obtained from quad sheets often are used to estimate the average project height above a datum to compute the sea level or ellipsoid reduction factor.

The size and type of the project, target scale, and contour interval will define the kind and accuracy requirements of the control to be established. After these have been established, instruments and the measurement system are selected. The measurement systems are

- triangulation
- trilateration
- traverse
- inertial surveying
- GPS surveying
- geodetic leveling
- photogrammetry/analytical control

Of the above measurement systems, USACE typically uses traverse or GPS for horizontal control.

After the information has been sorted and permission to enter the property has been granted, the survey party will go to the field and recover the monuments that will be of benefit to the project. At the same time, traverse stations are strategically set, and lines may be cleared under the conditions of the right-of-entry. Monuments set by different agencies should be identified and noted. Information supporting the unfamiliar monuments may be obtained from the respective agencies. Without this supporting information, the positions of these monuments may never be of value to the project. Unanticipated monuments that are found should be noted, because they may have value that was not apparent when the party first walked the project. The key is to locate anything that remotely resembles a monument in the field but to not hold this monument as equal weight with other known monuments in the traverse adjustments unless the history is known. Use the coordinate as a check unless no doubt exists about the position.

Plane coordinate systems can be used for large projects. The horizontal control is installed, measured, and adjusted by conventional traverse methods. This traverse is tied to a geodetic control network by including a geodetic monument (or monuments) in the traverse or by locating the geodetic monuments by ties (spur lines).

**Table 3-2. USACE Point Closure Standards for Vertical Control Surveys**

| USACE Classification | Point Closure Standard (ft) |
|---|---|
| Second-Order Class I | $0.025 \times M^{0.5}$ |
| Second-Order Class II | $0.035 \times M^{0.5}$ |
| Third-Order | $0.050 \times M^{0.5}$ |
| Fourth-Order—Construction Layout | $0.100 \times M^{0.5}$ |

*Note:* $M^{0.5}$ is the square root of distance $M$ in miles.

## 3-4. Primary Survey Control

Primary project control is set to establish survey control over large areas. USACE primary project control may be geodetic or SPCS. Geodetic primary project control may be obtained from the National Geodetic Survey (NGS). A vast amount of high-order survey information is compiled and adjusted by this organization to provide horizontal control and benchmarks in the United States. The latest adjustment of these data, including Canadian and Mexican geodetic survey data, establishes the vertical and horizontal datums for the North American continent. The newest vertical datum is the North American Vertical Datum 1988 (NAVD 88). The latest horizontal datum is the NAD 83. Connections with the NGRS shall be subordinate to the requirements for local project control.

## 3-5. GPS Survey Control

New projects may exist in areas that have no primary control available. GPS surveying has proven highly effective in transferring control from monuments outside the project. Differential GPS applications are capable of establishing control to accuracies better than 1:100,000. Although GPS is referenced to World Geodetic System of 1984 (WGS 84) rather than Geodetic Reference System of 1980 (GRS 80; ellipsoid definition used in NAD 83), the difference in resulting coordinates obtained through differential applications is negligible. Therefore, for differential applications, GPS can be assumed to be a NAD 83-referenced system. For details of GPS operation, see EM 1110-1-1003.

## 3-6. Secondary Control for Topographic Surveys

Secondary control traverses are constructed as closed loop traverses, closed traverses, or nets on flat plane coordinate systems. Loops can be adjusted by compass rule or least squares. Nets are best adjusted by least squares. Nets evolve in projects from a closed loop traverse where cross-tie traverses are later tied across the project. Reductions are made to survey data to project the control to the reference vertical datum. Another reduction for state plane coordinates places the control on the flat plane projection. Two points on the surface of the Earth are really separated by an arc distance. If the Earth radius varies in length, then so varies the distance between the same two points in a direct relationship. Following these two reductions, any distance between the control points will appear to be in error until the distance is divided by the combination of the two scale factors, called the combined scale factor. A local secondary system avoids this multiplication/division exercise to allow the surveyor to make environmental corrections to measurements without additional conversions. Later, the entire project will be converted from local to datum plane after field operations have ceased. Horizontal control for topographic surveys and construction surveys is designed primarily to provide positions within allowable limits. With the electronic equipment used today, the largest kind of error is the blunder. Systematic error can be reduced through calibration checks. Random error is more significant for primary survey control and stakeouts of large expensive structures. For topographic control, angles should be doubled and distances should be recorded ahead (AHD) and back (BK) at each traverse station. Thus, each measurement is recorded twice. This procedure is only recommended, not required, to avoid blunders. The traverse is adjusted, and a relative error of closure is obtained. This relative position closure provides a statement of protection against blunders and possibly some systematic errors. The error of closure of this control should be slightly less than the primary control used to set up this third- or fourth-order traverse. As long as the points located from this control can meet the requirements (including points far from the control but still within the map sheet), then the control is acceptable. This control is considered to be independent of the control used to check the topographic map.

## 3-7. Plane Coordinate Systems

If the project is 6 mi or less, then a flat plane can represent all the control points, although all USACE projects should be on an SPCS or a UTM system. The only exception is a project datum (local plane) system used for construction.

Each state in the United States has state plane grids that stretch 158 mi or less in a flat plane. Two kinds of map projections are typically used to project the curved Earth surface onto a flat plane. States elongated in the north–south direction typically adopt a Transverse Mercator (TM) System; 18 states use this system exclusively. States that are elongated in an east–west direction use the Lambert System; 31 states use this system. The states of Alaska, Florida, and New York use both systems. Also, Alaska uses an oblique Mercator projection on the southeast panhandle.

Another plane coordinate system used by the military or on large civil projects is the UTM coordinate system. The UTM system is divided into 60 longitudinal zones. Each zone is 6° in width, extending

3° on each side of the central meridian. The UTM system is applicable between latitudes 84° N to 80° S.

For first- or second-order surveys, all distances will be reduced by the first two of the following corrections:

- reduction to reference elevation datum
- scale factor
- curvature correction

The horizontal curvature correction is used for precise surveys where distances are measured in excess of 12 km (7.5 mi). The arc/chord distance change is <1 part per million (ppm) and is not significant for most USACE applications. Note: The curvature correction, if necessary, is applied to field-measured distances before any other reductions or adjustments are performed. The combined scale factor will be applied to the rectangular coordinates if only one point is held fixed for the traverse adjustment.

The vertical curvature correction may be significant for precise surveys that use electronic equipment. Optical levels used for ordinary survey work should not be affected by curvature, because a balanced shot to either the backsight or the foresight should not exceed 300 ft.

The first two corrections listed above, reference elevation reduction factor and scale factor, often are combined in practice. The product of these factors is called the "combined factor." Survey parties using primary control for construction stakeout purposes need the combined scale factor listed on the particular construction drawing sheet being used for the stakeout. The survey party divides inversed grid control coordinates by this correction to check measurements between the recovered traverse stations. If the field measurements check, the stakeout proceeds.

If any of these corrections are applied to survey measurements, a formal note must be printed on all drawings depicting the horizontal control. This note shall be the link between the points on the ground surface and the plane coordinate system used for the project.

## 3-8. Scale Factor Considerations

USACE projects typically use SPCS projections; therefore, coordinate problems may be present in high elevation locations or route surveys that extend for long distances.

### a. Projects in High Elevations

For projects in high elevations mapped to large scale, a project datum may be necessary to separate the survey measurements at ground scale from coordinates reduced down to the SPCS projection. This project datum eliminates blunders. The term "project datum" is used in lieu of the term "local datum" to mean the project datum is not translated or rotated from the SPCS orientation. Local datums in USACE can be rotated to station numbers and offsets from center lines.

### b. Constructing a Project Datum

To construct a project datum, a digital computer file of SPCS coordinates is typically used. A large constant (number) is subtracted from every northing and easting to identify the coordinates as entirely different from the SPCS coordinates. Next, all northings and eastings are divided by the selected combined scale factor to put the coordinates at the ground scale. The project datum is now established.

### c. Project Datum Example

For a project datum example, Figure 3-1 can be used to illustrate the procedure. The traverse stations labeled A1, A2, A3, ..., C1, C2, C3 may be used to establish the project secondary control. The digital computer file of project secondary coordinates is used to establish a project datum for field use. Say 20,000,000 was subtracted from all the northings and eastings to alert field crews that this coordinate file is not on the SPCS. Next, the combined scale factor of 0.994632 from Figure 3-1 is divided into all northings and eastings. Now, field crews can operate in a project datum without confusion about control distances that don't match ground distances. A note shall be added relating the project datum back to the SPCS. The coordinates shall be titled "Project Datum Coordinates." Figure 3-2 is a worksheet for use in converting state plane coordinates to project datum coordinates.

### d. Reduction to Sea Level

Distances between geodetic monuments are reduced to sea level for the North American Datum of 1927 (NAD 27). Sea level was the intended datum for North America, modeled from the Clark 1866 spheroid. The vertical datum used relative to 26 primary tide station local sea levels is the National Geodetic Vertical Datum of 1929 (NGVD 29). Distances between geodetic monuments of NAD 83 are reduced to the ellipsoid. The ratio used to compute this scale factor is

base length/ground distance = $R/(R + H + N)$

where

$R$ = mean Earth radius (20,906,000 ft)

*Example:* A distance of 1,000.00 ft is measured between monuments where

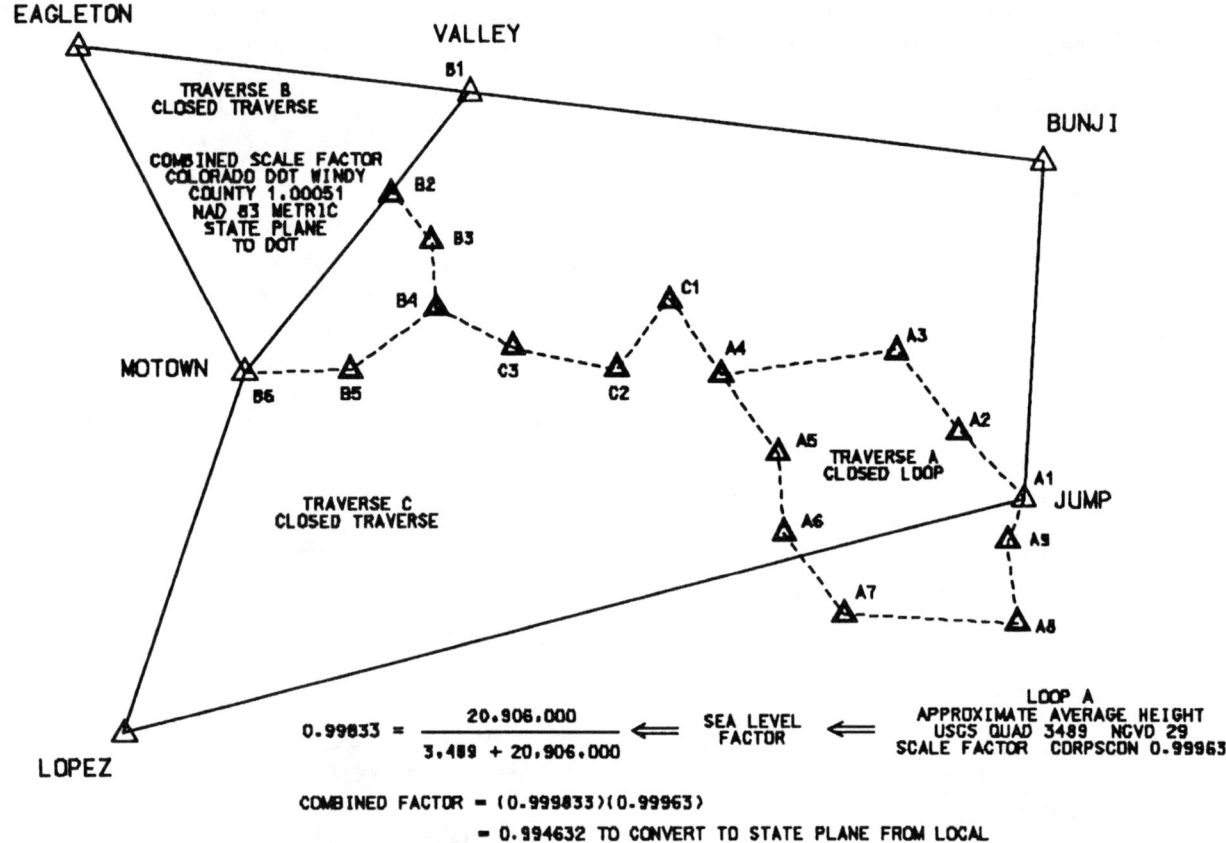

Figure 3-1. Primary, Secondary Project Control

mean height (H) = 2,416.1 ft

mean geoid height (N) = –58.4 ft

ellipsoid factor = $R/(R + H + N)$

= 20,906,000 / (20,906,000 + 2,416.1 + –58.4)

= 20,906,000 / 20,908,357.7

= 0.99988724

Therefore,

base length = ground distance × ellipsoid factor
= 1000 ft × 0.99988724
= 999.89 ft

Scale is the grid distance divided by the geodetic distance.

scale factor = grid distance/ground distance

The grid distance is the scale factor times the geodetic distance. Scale factors may be less than, equal to, or greater than 1.0. Lambert Conic projections vary scale in the north–south direction. UTM and TM projections vary in the east–west direction.

## 3-9. Control Checks

If the existing control is geodetic, the lengths between the monuments will have been reduced to reference vertical datum. The azimuth between them will be a geodetic azimuth and any reference monuments in the vicinity will have geodetic azimuths. Field checks will require corrections to the measurements in many cases. For high elevations, distances could measure longer by a tenth or more in 1,000 ft.

### a. Coordinate Transformations

Differential GPS units can measure the distance between two points on the basis of the datum selected in the GPS receiver unit. A three-dimensional vector is established between the master station and the remote. The initial computation in both receivers is based on the WGS-84 ellipsoid and the Cartesian vector relationship from the Earth-center to each unit. Vector subtraction between the Earth-center vectors provides the WGS 84 Cartesian vector between the units. Any errors are assumed to be sub-

# TOPOGRAPHIC SURVEY CONTROL

| Project Name | Zone (North or South) | Minimum Latitude | Maximum Latitude | Mean Latitude | Mean Elevation | Mean Geoid Height |
|---|---|---|---|---|---|---|
| (SF) Obtained by using the mean latitude from the table k value | (EF) $EF = R/(R + N + H)$ where $R$ = assumed radius of Earth of 6,372,000 (m) $H$ = mean elevation from sea level for project (m) $N$ = mean geoid height for project (m) | | (CF) $CF = SF \times EF$ | | (N) and (E) Project Datum $N = (Y/CF) + 1{,}000$ $E = (X/CF) + 1{,}000$ | Conversion to US Survey Foot $N \times (39.37/12)$ $E \times (39.37/12)$ |

| Point ID | (Y) State Plane Northing (m) | (X) State Plane Easting (m) | (SF) Mean Scale Factor (8 significant figures) | (EF) Mean Elevation Factor (8 significant figures) | (CF) Combined Factor (8 significant figures) | (N) Project Datum Northing (m) | (E) Project Datum Easting (m) | Project Datum Northing (ft) | Project Datum Easting (ft) |
|---|---|---|---|---|---|---|---|---|---|
| | | | | | | | | | |
| | | | | | | | | | |
| | | | | | | | | | |
| | | | | | | | | | |
| | | | | | | | | | |
| | | | | | | | | | |
| | | | | | | | | | |
| | | | | | | | | | |
| | | | | | | | | | |
| | | | | | | | | | |
| | | | | | | | | | |
| | | | | | | | | | |

Calculated By: _____   Checked By: _____

**Figure 3-2. Worksheet for Converting State Plane Coordinates to Project Datum**

tracted out during this process. Additional coordinate transformations provide the geographic coordinate based on the map projection parameters.

### b. Vertical Component

The vertical component of a GPS position is usually converted to a vector component in the normal direction to the surface at the receiver. The surface is the ellipsoid, not the geoid. At any point, the ellipsoid will generally be different than the geoid (mean sea level with undulations) or the elevation where the measurements are taken.

### c. Closed Loop Traverse

The first step in checking a closed traverse is the addition of all angles. Interior angles are added and compared with $(n - 2) \times 180°$. Exterior angles are added and compared with $(n + 2) \times 180°$. Deflection angle traverses are algebraically added and compared with $360°$. The allowable misclosure depends on the instrument, the number of traverse stations, and the intention for the control survey.

$$c = K \times n^{0.5}$$

where

- $c$ = allowable misclosure
- $K$ = fraction of the least count of the instrument, dependent on the number of repetitions and accuracy desired (typically 30" for third-order and 60" for fourth-order)
- $n$ = number of angles

Given the parameters, exceeding this value *may* indicate that some other errors, of angular type, are present in addition to the random error. The angular error is distributed in a manner suited to the party chief before the adjustment of latitude and departures. Adjustment of latitudes and departures is the accepted method in the USACE. If GPS is used, latitude, departures, and adjusted line lengths are computed after adjustment to obtain the error of closure and the relative point closure. The relative point closure is obtained by dividing the error of closure ($E_C$) by the line lengths.

relative point closure = $E_C / \Sigma$ of the distances

### d. Closed Traverse between Two Known Control Points

To establish a solid field technique, the initial azimuth shall be checked in the field with GPS or astronomic observations before this kind of traverse is continued. The extra time in minutes will save many work-hours of recomputation if the beginning or ending azimuth is not in the same meridian alignment.

The surveyor may decide to hold both ends of the traverse as fixed and adjust only the measurements in between. The misclosure between the two fixed points provides the expected error at the distance from the first control point. The surveyor holds one control point fixed and proportionally spreads (prorates) error between the points into the cross traverse as a fraction of the distance between the control points. The assumed error of the misclosure between the control points must be within the allowable limits of error for this procedure. For example, if control is in state plane coordinates (third order, class I), then the closure must be at least 1:10,000. If a traverse has legs that add up to 8764.89 ft, then

$$8{,}765\text{-ft length} \times (1\text{-ft } E_C)/(10{,}000\text{-ft length})$$
$$= 0.88 \text{ ft of allowable } E_C$$

The procedure for adjusting this kind of traverse begins with angular error just as in a loop traverse. To determine the angular error a formula is used to generalize the conversion of angles into azimuth. The formula takes out the reciprocal azimuth used in the backsight as $(n - 1)$ stations used the back-azimuth as a backsight in recording the angles.

$$A_1 + \alpha_1 + \alpha_2 + \alpha_3 + \ldots + \alpha_n - (n - 1)(180°) = A_2$$

If the misclosure is exceeded, the angular error may have been exceeded, or the beginning and ending azimuths are in error or oriented in different meridian alignments.

GPS points can aid the closure process by establishing point pairs at the endpoints with conventional surveys between the point pairs. Do not exceed 15 km in distance from the master station to the point pairs. Also, separate the point pairs by a distance of at least 400 m to make the constant GPS error (cm ± ppm) insignificant. If GPS is not available, sun shots can be taken at both ends of the traverse. The sun shot will establish an azimuth reference only where GPS provides azimuth and position with point pairs.

If beginning and ending azimuths were taken from two traverses and the angle repetitions were found to be at least an order of magnitude better than the tabulated angular error, the ending azimuths may contain a constant error that may be removed to improve the allowable error. GPS or astronomic observations may be used to find the discrepancy if the benefit of this procedure is worth the extra cost of the improved traverse to the total project costs.

The beginning and ending azimuths are measured and converted to grid. The check grid azimuth may then be compared with the grid azimuth of the traverse (if the traverse was in grid).

$$\varepsilon = \text{measured} - \text{true}$$

$\Delta\alpha$ = recorded azimuth − GPS azimuth

$\Delta\alpha_{BK} = \Delta\alpha_{AHD}$

$\Delta\alpha_{BK} - \Delta\alpha_{AHD} = 0$

true = measured − $\varepsilon$

The error is algebraically subtracted from the angle misclosure. The absolute value of the misclosure should be less than the initial misclosure because of external plus internal error. This new misclosure is checked against the allowable error. If the misclosure is now acceptable, proceed with adjustment. The angular misclosure is spread as a cumulative azimuth error from the beginning or ending station. For example, if the angular misclosure was 2 minutes in excess of the above formula, and nine angles were recorded, then

2 minutes × (60 seconds/1 minute) × (1/9) angles = 13.33 seconds per station

subtract (13.3)(1) = 13 seconds from azimuth 1

subtract (13.3)(2) = 27 seconds from azimuth 2

subtract (13.3)(3) = 40 seconds from azimuth 3

$\vdots$

subtract (13.3)(9) = 120 seconds = 2 minutes from azimuth 9

Note: The significance of seconds depends on the least count of the instrument.

# CHAPTER 4

# TOPOGRAPHIC SURVEY TECHNIQUES

## 4-1. General

This chapter outlines the most common field techniques used in performing topographic surveys. The primary focus is on electronic plane-table methods and electronic total station techniques. Transit station methods are rarely used, and they are not covered. Photogrammetric methods of acquiring topographic data are covered in EM 1110-1-1000. Kinematic Global Positioning System (GPS) topographic surveying methods are covered in EM 1110-1-1003. GPS and total station equipment usage are the best combination for topographic map production. This chapter focuses on topographic surveying techniques used for detailing large-scale site plan maps for engineering and construction.

## 4-2. Engineering Site Plan Surveys

An engineering site plan survey is a topographic (and, if necessary, hydrographic) survey from which a project is conceived, justified, designed, and built. The methods used in performing an engineering survey can and sometimes will involve all of the equipment and techniques available. Photogrammetry may be used to produce maps of almost any scale and corresponding contour interval. Profiles and cross sections may also be obtained from aerial photos. The accuracy of the photogrammetric product varies with the contour interval and must be considered when planning such a project.

The plane table and alidade may be used to produce maps in the field. A blank map on which control points and grid ticks have been plotted is mounted on the plane table. The table is mounted on a low tripod with a specially made head. The head swivels so that it can be leveled, locked in the level position, and then be rotated so that the base map can be oriented. The base map is a scaled plot of the ground control stations. Thus, with the table set up over one of the stations, it can be rotated so that the plotted stations lie in their true orientation relative to the points on the ground. Spot elevations and located features are usually located with an alidade that uses stadia to determine distance. The error of a map produced with a plane table and alidade varies across the map as the error in stadia measurements varies with distance. Horizontal errors may range from 0.2 ft at 300 ft to 10 ft or more at 1,000 ft. Because the elevation of the point is determined from the stadia measurement, relative errors in the vertical result.

Transit tape topographic surveys can be used to locate points from which a map may be drawn. The method generally requires that all observed data be recorded in a field book and the map plotted in the office. Angles from a known station are measured from another known station or azimuth mark to the point to be located and the distance from the instrument to the point. The elevation of the point is determined by stadia or other means such as chaining or electronic distance measurement (EDM) and differential leveling. The accuracy may be slightly better than the plane table/alidade method or very high (0.1 ft or less), depending on the equipment combinations used. Transit tape topographic survey methods are rarely used and are not covered in this manual. Procedures for performing transit tape topographic surveys can be found in older survey texts.

If there is extreme congestion, the plane table is often used in combination with transit tape. In this case, points are plotted as they are located. The advantage of the plane table, namely the developing map, at the job site does away with the disadvantage of transit tape (angle distance)—the absence of the real view while the map is being produced. The accuracy of the combined method is usually very high because equipment combinations are generally used that yield the best results. Often theodolite, EDM, and direct leveling are used to locate the point.

The last and most commonly used method is the total station. The remainder of this chapter discusses field topographic survey techniques using plane table and electronic total station methods.

## 4-3. Utility Surveys

### a. Definition
Utility surveys can be of several types, but principally they can be divided into two major types. One type is the layout of new systems, and the other type is the location of existing systems. The manual confines the definition of utilities to mean communications lines, electrical lines, and buried pipe systems (including gas, sewer, and water lines). The layout of new systems can be described as a specialized type of route surveying, in that they have alignment and profiles and rights-of-ways similar to roads, railroads, canals, etc. In reality, utilities are transportation systems in their own right. Utilities are special in that they may have problems regarding right-of-way above or below ground. This section does not cover route layout for utilities because route surveying principles are covered in a subsequent chapter.

### b. Uses
A great portion of utility surveying is the location of existing utilities for construction planning, facility alteration, road relocations, and other similar projects. This is a very important part of the preliminary surveys necessary for most of these projects.

### c. Techniques
Utilities are usually located for record by tying in their location to a baseline or control traverse. It may be more convenient to locate them with relation to an existing structure, perhaps the one that they serve.

Aboveground utilities are usually easily spotted and are a bit easier to locate than the buried variety. Therefore, they should present little difficulty in being tied to existing surveys. Pole lines are easy to spot and tie in. Consulting with local utility companies before the survey has begun will save much work in the long run. Any plats, plans, maps, and diagrams that can be assembled will make the work easier. If all else fails, the memory of a resident or nearby interested party can be of great help.

Proper identification of utilities sometimes takes an expert, particularly regarding buried pipes. There are many kinds of wire lines on poles in this modern age of electronics; this also can lead to identification problems. Where once only power and telephone lines were of concern, now cable TV, burglar alarms, and maybe even other wire line types must be considered. Much resourcefulness is required to identify the modern maze of utilities.

The location of underground utilities by digging or probing should be undertaken only as a last resort, and then only with the approval and supervision of the company involved. Some techniques that work are the use of a magnetic locator, a dip needle, or even "witching" for pipes or lines underground.

## 4-4. As-Built Surveys

### a. Definition
As-built surveys are surveys compiled to show actual condition of completed projects as they exist for record purposes and/or payment. Because many field changes—both authorized and clandestine—occur during construction, surveys are regularly completed to check the project against plans and specifications.

### b. Requirements
As-built surveys are usually a modified version of the preliminary survey that was originally required to plan the project. This is particularly true of road, railroad, or watercourse relocation projects. These projects are all of the route survey nature. The as-built survey, out of necessity, is also this type. The following items are typically checked:

- alignment
- profile or grade
- location of drainage structures
- correct dimensions of structures
- orientation of features
- earthwork quantities, occasionally

### c. Methodology
For route survey–type projects a traverse is usually run, and major features of the curve alignment are checked. Profiles may be run with particular attention paid to sags on paved roads or other areas where exact grades are critical. Major features of road projects that are often changed in the field and will require close attention are drainage structures. It is not uncommon for quickly changing streams to require modification of culvert design. Therefore, culvert and pipe checks are critical. The following items typically should be checked for all major drainage structures:

- size (culverts may not be square)
- skew angle (several systems in use)
- type or nomenclature (possibly changed from plans)
- flow line elevations (very important and should be accurately checked)

- station location of structure center line with regard to traverse line (should be carefully noted)

Utilities that have been relocated should be carefully checked for compliance with plans and specifications. Incorrect identification of various pipes, tiles, and tubes can result in difficulties. Because the subject is somewhat complicated, it is important to keep track of this kind of information for the future.

Project monumentation is sometimes a requirement of as-built surveys because it is common to monument traverse lines and baselines for major projects. Their locations should be checked for accuracy. In many areas, it is common for such monumentation to be done by maintenance people who are not at all familiar with surveying, and therefore, the work is not always as accurate as would be desired.

## Section I. Plane-Table Surveys

### 4-5. General

The Egyptians are said to have been the first to use a plane table to make large-scale accurate survey maps to represent natural features and manmade structures. Because of our familiarity with maps, it is easy for us to stand on a site with a scaled map, orient it in the direction of the features on the ground, and accurately picture ourselves on the map and on the ground itself. The survey map is the basic starting element of a civil engineering project. Plane-table surveying has, for most purposes, been replaced by aerial photogrammetry and total stations, but the map is still similar. The plane table is well-suited for irregular topography that exists in natural landscapes. Manmade symmetric landscapes such as cities and military installations are more suited to total-station or transit-tape methods. One advantage the plane table has over the transit-tape point locations is the accuracy of contour lines drawn by an experienced surveyor. The surveyor will interpolate enough controlling points to find the contours. Parts of contours are artistically drawn to capture the actual landscape without having to locate the focus of points that describe a trace contour. The topographic map is finished in the field by plane-table method. Digitizing plane-table sheets from experienced topographers provides more accurate contours than contours interpolated from point location. The plane-table method can also be used to spot-check existing maps with small scales (less than 1 in. = 100 ft). Allow for paper shrinkage due to ambient outdoor weather conditions. Plot digital maps on fresh Mylar to avoid this problem.

### 4-6. Plane-Table Topography

A topographic survey is made to determine the shape or relief of part of the Earth's surface and the location of natural and artificial objects thereon. The results of such a survey are shown on a topographic map. A plane table, telescopic alidade, and stadia rod are used to locate the required points, and they are plotted on the plane-table sheet as the measurements are made. The use of the plane-table method in obtaining topography affords the following advantages:

- The terrain being in view of the topographer reduces the possibility of missing important detail.
- Detail can be sketched in its proper position on the plane-table sheet from a minimum of measurements. This applies especially to the representation of relief where the contour lines can be sketched between plotted points.
- A greater area can be accurately mapped in a given period of time.
- Triangulation can be accomplished graphically to avoid office computation.
- Office work is reduced to a minimum.

### 4-7. Plane-Table Triangulation

Plane-table triangulation consists of the location of many points on a plane-table sheet through the use of the so-called pure plane-table methods. These involve the use of only the plane table and alidade, and in theory, the operations may be executed by one person. In practice, one or more systems are used, and auxiliary methods such as stadia are used. The area that is to be mapped is outlined on the plane-table sheet by means of a projection on which the initial horizontal control points have been plotted. Plane-table triangulation starts from these control points, and by means of a plane-table method of intersection and resection, the necessary points on which to tie the topographic details of the map are "cut-in" or located on the plane-table sheet.

#### a. Equipment

The plane table consists essentially of a drawing board that is supported by a tripod and used in connection with an alidade. The board can be leveled and also turned in any horizontal position and then clamped when properly set. When in use, the plane table and its support must never move. Once set or oriented, the greatest care must be taken that the position is not disturbed. The topographer must not lean on it or against it.

## b. Setup

After leveling the plane table, place the alidade on a line connecting the station occupied with one of the triangulation points farthest away. Revolve the table until the farthest signal is bisected by the vertical wire of the alidade, and clamp the table. Verify the orientation by sights to additional visible triangulation stations. Make the circuit of the horizon systematically, and take foresights to prominent objects, such as signals, towers, chimneys, flagpoles, monuments, church steeples, and definite points on schoolhouses, dwellings, barns, trees, or spurs. Draw the lines of sight with a chiseled-edge 9H pencil along the square edge of the alidade, being careful always to hold the pencil at the same angle and to see that the contact of rule and paper is perfect. Get azimuths of long straight stretches of roads and railroads whenever possible. Stadia may well be used to locate road forks or objects in the immediate vicinity of the station. From time to time while making observations and on the completion of the work at each station, check the orientation of the plane table to see whether there has been any movement.

## c. Vertical Angles

After all the sights have been taken, adjust the level of the alidade and read vertical angles to the points whose elevations are desired. Angles that are read to the principal control points in this scheme should be checked.

## d. Station elevation

The elevation of each plane-table station should be determined by means of vertical angles taken either to specifically located benchmarks or to other plane-table stations whose elevations have been determined previously. In general, the principal stations from which the greatest number of vertical angles are to be taken should be connected by means of reciprocal vertical angles taken under differing conditions. The final elevation of each station in the net should be determined by means of a weighted adjustment of the observed differences in elevation. In measuring important vertical angles, such as those to other stations or to points on level lines, all readings should be checked by reversing both level and telescope and by using different positions of the vertical arc. This can be accomplished by placing a plotting scale or some other flat object under one end of the alidade ruler.

## e. Records and Computations

Plane-table stations may be designated in the same manner as traverse stations. Such designations should be written on the plane-table sheet and also entered in a suitable notebook. A brief description of objects sighted may be noted on the plane-table sheet and written along the line of sight. Complete descriptions should be recorded in the notes. The notebook should also contain the vertical-angle records.

Distances between stations and located points whose differences in elevation are to be computed will be scaled by means of a boxwood scale provided for the purpose and graduated for the field scale. Computation of differences in elevation is facilitated by the use of prepared tables. Care should be exercised in noting and in allowing for the height above ground for both the alidade and the point sighted. The proper corrections for the refraction should be applied whenever appropriate.

When the work on the initial station is finished, repeat the operation on the station at the other end of the base and on as many additional stations as may be necessary to complete the work. If practical, all triangulation stations that have been plotted on the sheet should be occupied. The point of intersection of the lines drawn to same objects determines its location, but all intersections should be verified by a sight from a third position. Additional stations may be made at intersected points, at points to each of which a single foresight has been drawn on the plane-table sheet, and at points whose locations may be determined by the "three point method." Figure 4-1 illustrates this method.

## f. Suggestions

A signal should be erected where necessary to mark the place of a station for future reference. Care should be taken to prolong on the plane-table sheet a line that may later be used for an orientation. In areas of great relief and difficult access, every opportunity should be taken to contour topographic features. These include bottoms of canyons, rock exposures along canyon walls, and ground surfaces in heavily timbered areas.

## 4-8. Plane-Table Resection

The location of a plane-table station may be obtained by the method of resection (Figure 4-1),

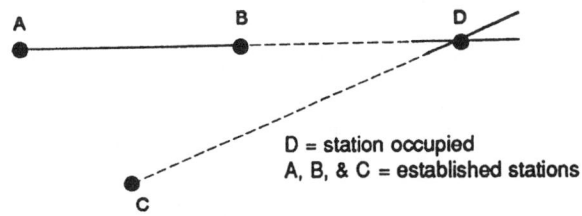

**Figure 4-1. Resection—Orientation by Backsighting**

which is stronger than the three-point method. Location by resection involves two separate operations performed on two different stations, and in practice, any length of time may lapse between the two operations. The resection method is of limited use because it involves a foresight from a previous station and the erection of a signal on the proposed station point or the positive identification afterward of a point or direction sighted. Foresight lines that are to be used for a location by resection should be drawn on the plane-table sheet exactly through the point that represents the station from which the sights are taken and to the full length of the alidade ruler. The line through the point representing the occupied station should be light, and care should be taken to hold the chiseled pencil point directly against the ruler. The foresight line need only be drawn through the approximate position of the new station and at the extreme end of the ruler. Doing so will avoid adding unnecessary lines to the sheet. After drawing the line, look through the telescope again and test the plane-table orientation to ensure that the table has not moved. Similar foresight should be taken when considering perspective station points whose locations may be obtained at some future time by resection.

To locate a new station that is being occupied and to which a long foresight has been drawn previously, orient the plane-table sheet by placing the alidade ruler in the reverse direction of the line sighted. Swing the plane-table board until the station from which the foresight is drawn is seen behind the center wire and clamp the plane-table board. The station being occupied is on this foresight line. The location is determined by resection from other plane-table stations whose directions are most nearly at right angles to the foresight line that they are to intersect and whose distance is less than that of the station from which the foresight line was drawn. Center the ruler on the plotted position of one of these stations. Swing the telescope until the signal mark at that station is behind the center wire, and draw a line against the ruler to intersect the foresight line. The intersection marks the location of the occupied station. The intersection should be checked by at least one other resection from another station. The two resected lines should cross the foresight line at the same point.

### 4-9. Plane-Table Two-Point Problem

In Figure 4-2, a and b are the plotted positions of visible inaccessible control stations A and B. It is desired to locate and orient the plane table at an unplotted station C. Use the following procedure:

- Select an auxiliary point D where the resection lines from A and B will give a strong intersection (>30°). D must be located with respect to C so that cuts on A and B from C and D will also give strong intersections.

- Set up the plane table at point D, and orient by eye or with a compass.

- Resect on A and B. The intersection of these resection lines is d', the tentative position D.

- Draw a ray from d' toward C. Plot the point c' on this ray at a distance from d' that corresponds to the estimated distance from D to C.

- Set up the plane table at point C and orient by a backsight on D. The error in this orientation is the same in magnitude and direction as it was at D.

- Sight on A and draw the ray through c', intersecting the line ad' at a'. In a like manner, sight B to obtain b'.

- Quadrilateral a'b'd'c' is similar to ABDC because the line a'b' will always be parallel to the line AB, the error in orientation will be indicated by the angle between line ab and a'b'. To correct the orientation, place the alidade on the line a'b', and sight on a distinctive distant point. Then, place the alidade on the line ab, and rotate the plane table to sight again on this point. The plane table is now oriented, and resection on A and B through points a and b establishes the position of the desired point C.

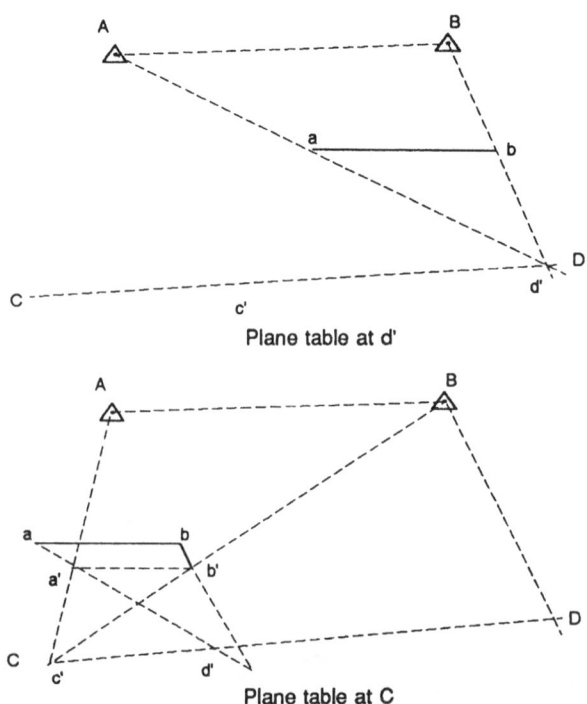

**Figure 4-2. Two-Point Problem**

## 4-10. Plane-Table Traverse

Traversing consists of much more than getting direction and distance, even though these are absolutely essential features. All essential topographic features on each side of the line are to be obtained at the time the traverse is made.

Accuracy of the plane-table traverse depends on two factors: obtaining and plotting of distances, and the orientation of the plane table.

Distances are obtained by stadia or tape, and the orientation is made by magnetic needle by backsight and foresight.

When the needle is used, the accuracy of the orientation depends on the freedom from local attraction and the length of the needle. For these reasons, it is necessary to avoid the use of the compass near railroads, electric transmission lines, or large bodies of steel or iron and in volcanic regions. No plotted line should be greater than the length of the needle.

The method used in determining distances will be governed by the character of the country and the scale of the work. Traverse lines should be run along roads, ridges, or streams or at intervals in timbered country when necessary. When the needle is used to set up alternate stations, intermediate stations should be used as turning points.

Streams near the roads should be mapped as accurately as the skill and experience of the topographer will permit.

When traversing railroads, frequent locations by the three-point method should be made whenever possible. The line should be extended by means of foresights and backsights. If this is not practical and it becomes necessary to rely on a needle, it is important to set up the plane table a sufficient distance from the rails to prevent their influence on the needle. The distances can be obtained advantageously by measuring a rail and counting the number of rails between stations.

To properly plot a new station, place the fractional scale division on the old point. Then prick the new station with a needle at the even division at the end of scale. This operation should be performed with great care, because more closure errors are attributed to careless plotting than to any other cause.

## 4-11. Plane-Table Stadia Traverse

When using plane table and stadia traverse, instrumental measurement of the distances in elevation gives sufficient control to permit considerable sketching to be done on either side of the line. Wherever possible, as in regions of low relief, elevations should be determined by using the alidade as a level and a rod as a level rod.

## 4-12. Plane-Table Three-Point Orientation

The three-point method involves orienting the plane table and plotting a station when three stations (established control) can be seen but not conveniently occupied. The following procedures should be used:

- Set up the plane table at the unknown point P (Figure 4-3) and orient by eye or compass.

- Draw rays to the known points A, B, and C. These rays intersect at three points (ab, bc, and ac), forming a triangle known as the *triangle of error*. The point ab denotes the intersect of the ray to A with the ray to B. Points bc and ac are similar in their notation. If the plane table is oriented properly, the rays to the three known points will intersect at a single point P rather than three points found in a above. To accomplish this result, turn the plane table several degrees in azimuth, then construct a second triangle of error (a'b', b'c', and a'c').

- Construct a third triangle of error (a"b", b"c", a"c"). A circular arc through points ab, a"b", and a'b' will pass through the point P; a circular arc through bc, b"c", and b'c' will pass through point P. These arcs should be sketched and the point P plotted.

- Orient the plane table by sighting from the plotted position of P, to the plotted position of any of the three known points.

- To check the solution, a fourth known station is observed when possible after completion of the three-point problem. If the ray through this fourth station does not intersect the other three rays at P, an error has been made.

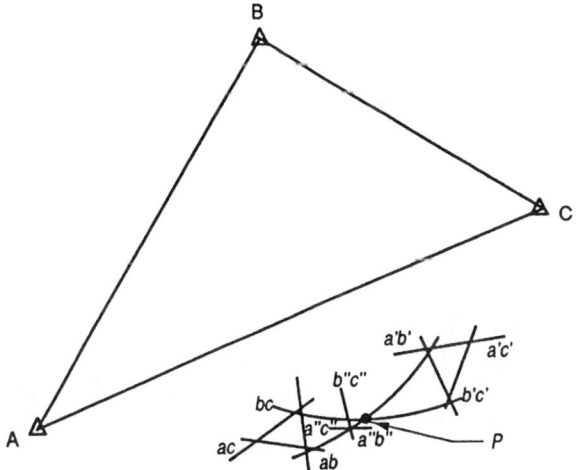

**Figure 4-3. Three-Point Problem**

## 4-13. Contouring Methods

Contours may be mapped from plane-table setups that are made directly over or adjacent to the country that is to be mapped. These areas include a traversed road, a table that is kept stationary while it is circled by one or more moving rods, or from a plane-table station that overlooks the distant country to be contoured. These three methods of contouring are described below.

### a. Contouring from a Traversed Line

In regions where the principal control is obtained by different kinds of traverse, the usual procedure is to first plot the contour crossings and other contours on or near the traversed road or other traverse lines. Then, extend the contouring out on both sides of the traverse line as far as good visibility and locally established control warrant. When the visibility from a traverse line is poor, points off the line may be used. Contouring from a traverse line also may be advantageously supplemented by having the rod held occasionally at salient points in the topography opposite the line. When all signs of a road circuit or other large traverse circuits have been traversed and mapped, traverses should be extended into or across the unmapped interiors, giving sufficient traverse control to enable the topographer to complete the mapping. Such interior traverses may be run across open country or along ridges or streams. The topographer may occupy favorable interior viewpoints with the plane table and reset from a previous traverse location to obtain a plane-table location.

### b. Contouring in Open Country

In open country of low relief, where little contouring can be done from single plane-table setups, one or two rodmen and a recorder can be used advantageously. When two rodmen are used, each holds a rod on different sides of the plane table and at salient points in the topography. Each rodperson advances in the direction of the proposed mapping. As soon as the sites become too long or are about to be obscured, both rodmen should hold their stations, and these points should be used as tuning points in the line. The mean of the two readings should be used in determining the elevation of the new plane-table station that is made beyond the points held by the two rodmen. The rodmen then advance as before. Plane-table locations may be obtained as in contouring, from a traverse line above or from stations below. To use this method effectively, the topographer should use signals between the table and the rod and between rods. The topographer should fully instruct the rodmen in their duties, because much depends on their activity and resourcefulness. The readings that result from successive rod sites may be plotted as they are taken, or depending on local conditions, they may be allowed to accumulate and plotted after the series has been completed.

### c. Contouring from Station

In open country of bold relief, where all the features are plainly visible, contours can best be delineated from plane-table stations overlooking the country to be contoured without the running of a traverse line or the use of a rod. The method of contouring from plane-table stations involves the use of plane-table triangulation and the three-point method. Woodland country as well as open country may be contoured from stations, provided a sufficient number of outlooks can be found from which a satisfactory view and a good determination of positions can be obtained. In the construction of contours from a station, the location and use of drainage lines is important.

Contouring from a station depends on supplemental control that is obtained by the location on the plane-table sheet by intersection method of many of the salient points on the surface that is to be contoured. For this reason, little sketching can be done from the first station other than form lines, which are later converted into placed contour lines. In planning the order of plane-table stations, careful consideration must be given to the need for the siting of many points ahead. A sufficient number of such sites may be intersected from subsequent stations and used as a basis for contour construction. Vertical angles may be taken from the points when they are first sited, after the points have been intersected, or at both times. In either procedure, the elevations must be computed and the contours placed on the map as soon as intersections are obtained.

The elevation of the plane-table stations must be determined from a carefully executed series of reciprocal vertical angles taken between the principal stations in the quadrangle. At least one station must be directly connected with a level benchmark by means of reciprocal vertical angles measured under different conditions.

### d. Contour Skeleton Outline

Before mapping the contours that are to represent the distance relief feature, a skeleton outline of that feature should be prepared. The quality with which this is accomplished depends on the accuracy, speed, and ease that the contours are placed on the map. Lacking an outline, an experienced topographer will make a suitable skeleton outline of the drainage and ridge lines before attempting the construction of the contour lines.

The landscape that is to be contoured should be divided into its separate features or unit masses, such as mountains, hills, and spurs. After sufficient control has been established through intersection methods, each feature segregated should have its natural drainage lines and boundaries sketched in. They should include tangents drawn to the salient points as well as located points being used as control for the placement of the drainage lines. Similarly, ridge and crest lines may be outlined. It is best to use convex forms as unit masses, such as spurs and lateral ridges. The intermediate drainage lines should be used as boundaries.

In determining elevations based on vertical angles, remember that large angles must be supplemented with accurate measurements of distance. Small angles based on measured distances that are approximate will yield useful elevations only for contour work.

Each separate unit mass should be as completely contoured as control and visibility permit before the contouring of another feature is begun. Should control alone be lacking, form lines should be lightly sketched in and advantage taken of a favorable view point. An effort should be made to cut-in the lacking control. By treating each mass as a separate unit, each can be delineated with its own characteristic shape.

## 4-14. Locating and Plotting Detail

Detail points are plotted to scale on the plane-table sheet with respect to the plotted position of the occupied station. Detail points are normally located and plotted by the radiation method, and instances are measured by stadia. Rays and instances are plotted directly on the plane-table sheet. The ground elevation at each detail point is determined, and the elevation is shown on the plane-table sheet beside the plotted position for the point.

Features shown on the map for which detail points must be located include the following:

- man-made works, such as buildings, roads, ridges, dams, and canals
- natural features, such as streams, lakes, edges of wooded areas, and isolated trees
- relief

On large-scale maps, it is often possible to represent the true shape of features to scale. On small-scale maps, buildings and other features must often be symbolized with the symbols centered on the true position but drawn larger than the scale of the map. Such detail is portrayed on the map by means of standardized topographic symbols.

Detail points and elevations for contouring are usually located at key points of distinct changes in ground slope or in the direction of a contour. Such key points are located at the following positions:

- hill or mountain tops
- on ridge lines
- along the top and foot of steep slopes
- in valleys and along streams
- in saddles between hills

Contour lines are drawn on the map by logical contouring. Ground elevations are determined at key points of the terrain, and these positions are plotted on the plane-table sheet. After a number of key points have been located (usually from an occupied station) and plotted, sketching of the contour lines is started.

Figure 4-4 shows a portion of a plane-table sheet on which some contour lines have been drawn. Key points are shown in this figure, with the interpolated positions of the contour lines plotted. Contours should be drawn as plane tabling progresses so that the topographer can use the view of the terrain when drawing the contours.

When contouring, it must be remembered that stream and ridge lines have a primary influence on the direction of the contour lines. Figure 4-5 shows several typical forms often encountered in contouring. It should be noted that the contour lines crossing a stream follow the general direction of the stream on both sides and cross the stream in a fairly sharp "V" that points upstream. Also, contour lines curve around the nose of ridges and cross ridge lines at approximate right angles.

Every fifth contour line should be drawn heavier than the others, and the elevations of these heavier lines should be shown at frequent intervals. These heavy, numbered contour lines are those representing multiples of the 5-, 10-, 25-, 50-, or 100-ft elevations. For example, with a 2-ft contour interval, the 10-ft contour lines would be drawn heavier and numbered.

Practical hints for plane-table mapping are

1. The topographer should face the area he is contouring.
2. Contour lines should not be drawn beyond the determined and plotted elevations.
3. The pencil should be kept sharp at all times.
4. Accurate plotting of distances is critical. For this reason, plane-table traverses should be kept to a minimum.
5. When drawing a ray toward a detail point, draw only a short line near the plotted position for the detail point.
6. The magnetic needle often can be used to orient the plane table approximately at the start of resection problem.

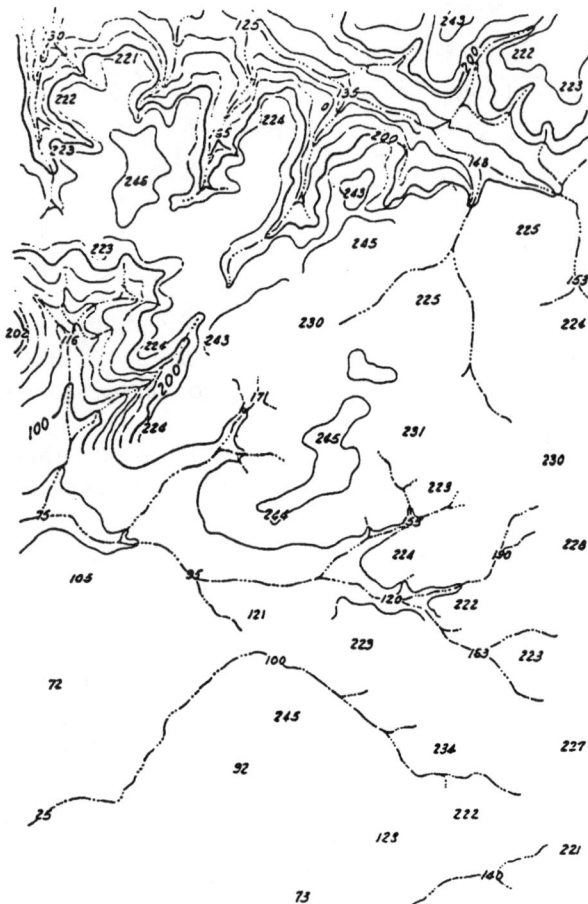

**Figure 4-4. Contouring**

7. The topographer should not remain at the plane table but move around enough to become thoroughly acquainted with the terrain.

## 4-15. Plane-Table Equipment Checklist

The equipment listed below should be on hand:

- alidade
- plane-table drawing board
- plane-table tripod
- thumb tacks (to secure map or plane-table sheet)
- sewing needles
- polymer 6J–9H pencil leads for Mylar
- 6H–9H pencil leads for vellum
- 3H–9H pencil leads for paper
- plumb bob
- cloth tape or reel chain
- pounce (powder to keep black marks off the map and help the straightedge to glide)
- 25-ft topo rod
- plane-table sheets

## 4-16. Plane-Table Setup Hints

Rotate the plotted point on the plane table over the point on the ground and in the direction of the backsight before completing the setup. Do this before placing the alidade on the table. The reason is the plotted point is generally not in the center of the plane-table board. Thus, any rotation about the Johnson head will send the plotted point away from the point on the ground. The wing nut for rotation is the lower nut or the one located farthest below the plane table.

Secure (snug, but not too tight) the Johnson head ball-and-socket wing nut for leveling (closest to the plane table). Make sure the plotted control point has been plumbed over the actual control point. Place the alidade in the center of the plane table or over the Johnson head. With one hand holding the plane table fixed, carefully loosen the leveling wing nut and level the table with the bull's eye bubble. Feel with the free hand and secure the upper wing nut (leveling nut). Check the setup for level and use a plumb bob to ensure the plotted point is over the actual point.

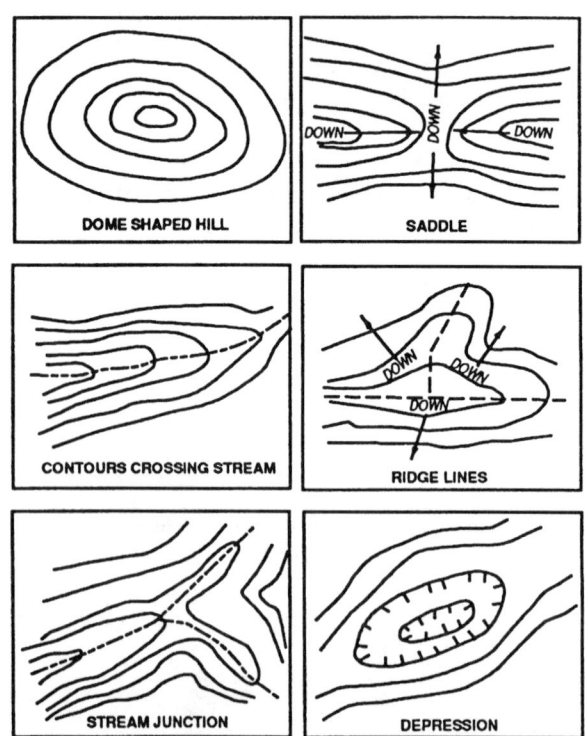

**Figure 4-5. Typical Topographic Forms**

## 4-17. Plane-Table Notekeeping

The following items must be recorded at every setup:

- Measure up from control point on the ground to the alidade trunion axis (height of instrument [HI] above ground). The alidade must be positioned over the point with the straightedge against the needle.
- No setup will exceed 0.1 ft in distance between the plotted control point plumbed over the actual control point when the station is being occupied as a setup.
- After setup initialization (table oriented and table level) a horizontal check on at least two points other than the point being occupied is necessary. One of these two points may be the backsight. Distance as well as alignment must be recorded to the accuracy (acceptable tolerance) of the survey in progress. If no other checkpoint is visible, a traverse must establish mapping control. This traverse can be performed with stadia using the plane table to mark the points. Two points other than the point occupied will also be used to check the vertical. A beamen arc will be used to check one of these elevations. Record all observations for checks neatly into the field book.
- The distance between plotted points on the plane-table sheet shall not exceed 1 in. For example, a plane-table survey to be performed at 1 in. = 20 ft shall have ground shots spaced no more than 20 ft apart.
- No point plotted on the plane-table sheet shall have a vertical error >0.3 ft.
- All manmade concrete structures will be located vertically to the hundredth of a foot.
- Hand-drawn contours shall be established by interpolation in two directions for every tick made on the plane-table sheet.

## 4-18. Plane-Table Location Details

The following details should be noted on the map:

- Railroads: Location of one track, track widths, number of tracks for parallel tracks
- Bridges: Bridge type, center lines, type of surface, number of lanes, bridge piers, fender systems, distance from center line to bottom of typical bridge beam, utility cables
- Tunnels: Center lines, approach islands, utility cables; develop a typical tube section.
- Airports: Center-line runways and magnetic azimuth of runway, runway thresholds, runway lights (amber), taxiways, taxiway lights (blue), possible obstructions to aircraft (tall trees, towers, buildings), control tower plus height, navigation aids (ILS localizer, VASI, VOR, VORTAC, NDB, LOM, RVR). Note: Control monuments are usually located in the area of the runway threshold.
- Dams: Dam type, center line, typical cross-section, toe of slope, water line
- Canals: Bulkheads, piers, pilings, dolphins, cable crossings, riprap or armor stone perimeters
- Roads: Number of lanes, type of surface, center lines, all edges of pavements, drainage ditches, driveway entrances, driveway culverts, road intersections, curb and gutter (vertically to 0.01 ft)
- Utilities: For underground utilities, call the proper agency for line location stakeout. Locate all stakes set in the field by this agency. Failure to locate all stakes could result in serious construction injuries.

    — Power poles: Pole numbers and agency names
    — Overhead powerlines
    — Overhead telephone lines
    — Overhead cable television lines
    — Storm sewers (curb inlets, yard inlets, etc.): Throat size, pipe sizes, inverts in and out
    — Sanitary sewers: Low side rim elevations and all inverts
    — Gas lines: Note diameters, tees.
    — Water lines: Note diameters, tees.
    — Fire hydrants: Never place a benchmark on the top nut; use a bell bolt.
    — Underground electric: Locate transformers, building connections.
    — Underground gas petroleum: Locate all tees and cutoffs.

- Buildings: Schools, churches, homes, industrial, office, barns, warehouses

    — Locate at least three corners and chain all distances completely around the perimeter of the structure. Include any overhangs.
    — Show overhead, ground surface, and underground utilities serving the building.
    — Locate any ruins of buildings.
    — Tanks (water, gas, oil, etc.)
    — Wells
    — Pumps
    — Septic fields, if possible
    — Fences

- Walls
- Sheds (note type of foundation)

- Cemeteries: Carefully locate all graves.
- Waterways

  Bulkheads
  Piers
  Dolphins
  Piles
  Docks
  Slips (concrete or wood)
  Buoys
  Oyster grounds
  Dikes
  Riprap (cobble size to armor stone size)

- Natural features

  Ridge lines
  Center-line swales
  Change in slope
  Valley lines
  Creeks (top of banks and center line)
  Springs
  Ponds
  Lakes
  Edges of woods
  Swamps
  Rock outcroppings
  Mines
  Trash dumps

# Section II. Electronic Total Station Surveys

## 4-19. Electronic Total Stations

Traditionally, surveying has used analog methods of recording data. The present trend is to introduce digital surveying equipment into the field. The fastest digital data collection methods are now done by electronic total stations. Total stations have dramatically increased the amount of topographic data that can be collected during a day. Figure 4-6 is a flow diagram of the digital information from field to finish. The method is well-suited for topographic surveys in urban landscapes and busy construction sites. Modern total stations are also programmed for construction stakeout and highway center-line surveys. Total stations have made trigonometric levels as accurate as many of the differential level techniques in areas possessing large relief landforms. These instruments can quickly transfer three-dimensional coordinates and are capable of storing unique mapping feature codes and other parameters that in the past could be recorded only on paper media, such as field books. One of the best features of the total station is the ability to electronically download data directly to a computer without human blunders. The use of electronic storage can result in a blunder of the worst magnitude, if reflector prism HIs are changed and not noted. Feature codes entered in error are also potential map errors that can be detrimental to an entire project.

### a. Total Station Surveys Are Like Plane-Table Surveys

The advent of the total survey station has made it possible to accurately gather enormous amounts of survey measurements quickly. Even though total stations have been around for more than 20 years, they are just now beginning to become popular among the general surveying and engineering community. In the past 10 years, total stations and data collectors have become common field equipment. As time progresses, they will increase in number and variety. The majority of survey firms are using data collectors today.

### b. Reasons for Implementation

In the early 1980s, the surveying instrument manufacturers introduced what has become a true total station, redefining the term by creating an entirely electronic instrument. In other words, the readout on the display panels and the readout from the EDM are in a digital form. This feature eliminated the reading errors that can occur with an optical theodolite. Also with the advent of the electronic theodolite came the electronic data collector, thus minimizing both the reading errors and the writing errors. Because the information in the data collector is interfaced directly to a computer, errors that occur in transferring the field information from the field book to the computer are eliminated. At this point, one can measure a distance to a suitable range with an accuracy of better than 5 mm plus 1 ppm, and angles can be turned with the accuracy of one-half arc second, all accomplished electronically. The vast increase in productivity that is available to us from what we had previous to 1980 is due to this modern equipment. In most land-surveying situations, the normal crew size can be reduced to two when equipped with an electronic theodolite. Because the data acquisition time is so fast, in some situations, three men are warranted

# TOPOGRAPHIC SURVEY TECHNIQUES

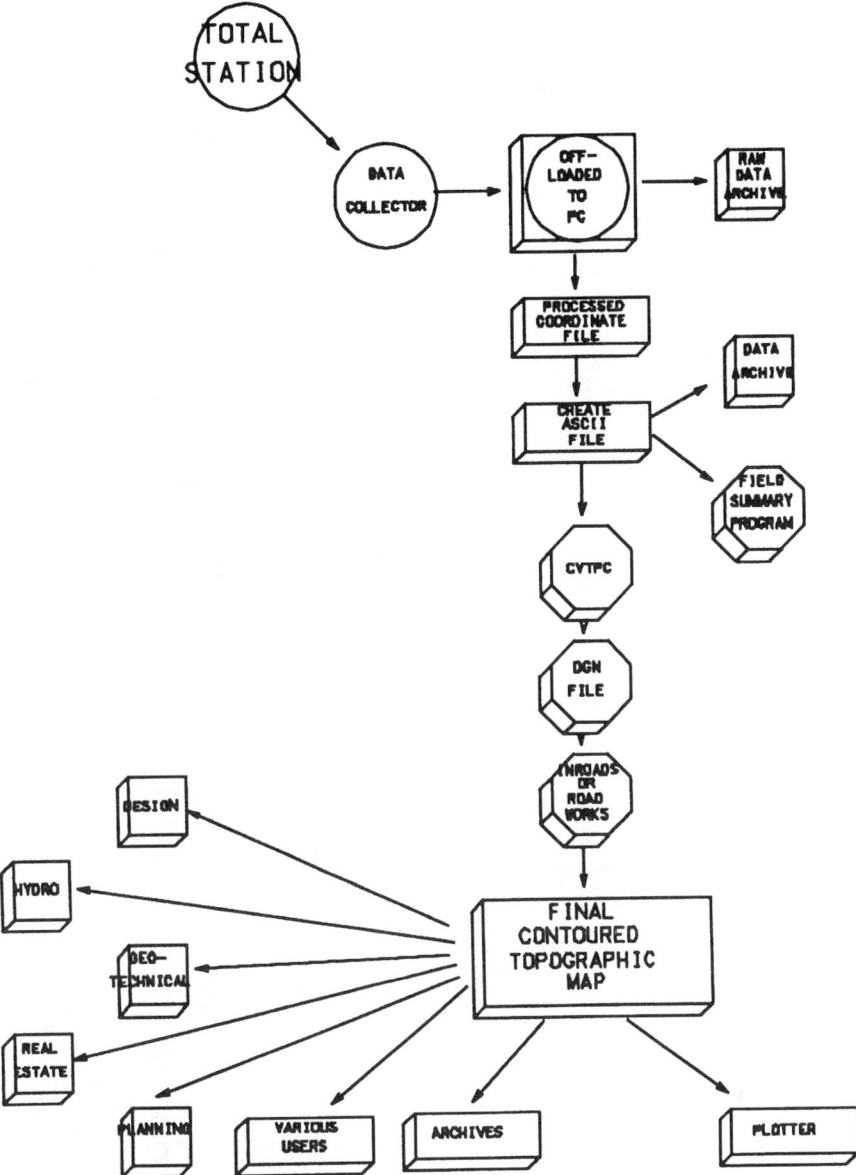

**Figure 4-6. Total Station Data—Field to Finish**

when it is possible to use two prism poles. This often results in an overall reduction in man-hours spent on the job.

### c. Field-Finish Surveying Capabilities

Figure 4-6 depicts the capability of an electronic total station to perform direct field-to-finish mapping products.

## 4-20. Field Equipment

Modern electronic survey equipment requires surveyors to be more maintenance conscious than they were in the past. They have to worry about power sources, downloading data, and the integrity of data as well as whether or not the instruments and accessories are accurately adjusted and in good repair. When setting up a crew to work with a total station and data collector, it is helpful to supply the party chief with a checklist to help the crew maintain its assigned equipment and handle the collected data on returning to the office. It is also important that each crew be supplied with all necessary equipment and supplies. These should be stored in an organized and easily accessible manner.

Let us first examine an equipment list that will assure the survey crew (two-person crew, consisting of a party chief/rodperson and a notekeeper/instrument person) a sufficient equipment inventory to meet the general needs of boundary, layout, and topographic surveys. Also, assume that this crew has a regular complement of supplies such as hammers, shovels, ribbon, stakes, lath, and safety equipment. This discussion will be confined to what is needed to maximize productivity when using a total station with data collector.

The minimum equipment inventory required is as follows:

- total station
- data collector
- batteries for 14 h of continuous operation
- tripods (3)
- tribrachs (3)
- target carriers (3)
- plumbing pole (1)
- target holders (4)
- reflectors (4)
- target plates (4)

With this equipment inventory, a two-person field crew will be able to handle most survey tasks that are routinely encountered in day-to-day operations. An additional tripod, plumbing pole, carrier, tribrach, and reflector would give the crew even greater flexibility and allow them to handle many projects more efficiently. It is also helpful for the field crew to have a convenient place to store their assigned equipment. Equip crews with briefcase-sized cases that will hold three tribrachs, four reflectors with holders, three carriers, and four target plates. A hard camera case or pistol case works well for this purpose. With all the components stored in one place, it makes inventory of the equipment easy and reduces the chance of equipment being left at the job site. This also allows for proper equipment maintenance.

## 4-21. Equipment Maintenance

The following checklist will aid each crew in properly maintaining and keeping inventory of its assigned equipment. At the end of each workday, the party chief should check that the following duties have been performed:

1. Clean all reflectors and holders. A cotton swab dipped in alcohol should be used on the glass surfaces. A crew member can do this during the trip back to the office.

2. Clean tribrachs. They should be dusted daily.

3. Remove dust from all instruments. A soft paintbrush or a shaving brush works well. If an instrument has been exposed to moisture, thoroughly dry it and store in an open case.

4. Download the data collector to the computer.

5. Backup all files generated from the download and check the integrity of the backup files before erasing the field data from the data collector.

6. Clean batteries and connect to charger. Some batteries require a 14- to 20-h charge, so one set of batteries may have to be charged while a second set is in operation.

## 4-22. Maintaining Battery Power

One of the biggest problems faced by the users of total stations with data collectors is maintaining an adequate power supply. Several factors should be considered when assessing power needs.

### a. Kind of Survey

A topographic survey entails much more data than a boundary survey. Normal production in a topographic mode is 200–275 measurements per day (350 shots per day is not uncommon). A boundary survey can entail making 16 measurements or so from each traverse point and occupying 10–15 points per day. Determine the number of measurements that you would normally make in a day, and consult the manufacturer's specifications to determine the number of shots you can expect from a fully charged new battery.

### b. Age of Batteries

Keep in mind that batteries will degrade over a period of time. This means that a new battery, with sufficient power for 500 measurements when new, may only be capable of 300 measurements after a year of use.

### c. Time Needed to Charge Batteries

Some batteries take up to 14 h to fully charge. If the work schedule will not permit 14 h for charging a second set of batteries, a battery with adequate power to supply the instrument for more than 1 day should be purchased.

### d. Power Requirements of Equipment

Older recording total stations that write data to tape will use up to three 7-amp/hour batteries per day. A newer instrument by the same manufacturer will take

3 days to use one battery. Because newer instruments use far less power than those on the market 6 years ago, this should be considered in determining power needs.

In addition to the proper assessment of power need, a record of the history and current status of the power supply should be readily available. When batteries begin to get weak, there is generally a rapid deterioration in their performance. To monitor the performance of a particular battery, record the serial number in a battery log book. If problems arise with a particular unit, check the log to see when the battery was purchased or when it was last recelled. Next, try discharging and recharging the battery. If performance is still not up to speed, have it checked to identify the weak cell and replace it. If the battery was not new or was recelled in the last year, recell the entire unit. When one cell goes, the next one is usually only a charge or two from failure. The cost of having a battery recelled is minimal when considering the cost of lost worktime due to power failure.

Also record the date the battery was charged on the shipping label that is attached to the battery box. When the battery is fully used, simply cross out the date, thus eliminating the confusion of not knowing which battery needs to be charged. Monitor the shelf time of the battery, and if it exceeds 10 days, recharge it. This keeps the power supply at peak performance. Always consult the operator's manual for recharge specifications.

It is always a good idea to have backup power available for that last 15 min of work. Most manufacturers can provide cabling for backup to an automobile battery. Some can even supply a quick charge system that plugs into the cigarette lighter.

Some power pointers include the following:

- Assess power needs for the particular job.
- Assess power usage of the equipment.
- Monitor performance of each battery.
- Monitor battery age, usage, and recell information.
- Have 1 day's worth of backup power readily available.

## 4-23. Total Station Job Planning and Estimating

An often-asked question when using a total station with a data collector is, "How do I estimate a project?" To answer this question, first examine the productivity standards expected of field crews.

Most crews will make and record 200–275 measurements per day. This includes any notes that must be put into the system to define what was measured. When creating productivity standards, keep in mind that a learning curve is involved. Usually it takes a crew four to five projects to become familiar with the equipment and the coding system to start reaching the potential productivity of the system.

A two-person crew is most efficient when the typical spacing of the measurements is less than 50 ft. When working within this distance, the average rodperson can acquire the next target during the time it takes the instrument operator to complete the measurement and input the codes to the data collector. The instrument operator usually spends 20 sec (more or less) sighting a target and recording a measurement, and another 5–10 sec coding the measurement.

When the general spacing of the data exceeds 50 ft, having a second rodperson will significantly increase productivity. A second rodperson allows the crew to have a target available for measurement while the first rodperson is moving. If the distance of the move is 50 ft or greater, the instrument will be idle with only one rodperson.

When dealing with strip topographic situations, data must be acquired every 3 ft along the length of the job. In urban areas, the data may seem to be more dense, but the rights-of-way are generally wider. The rule-of-thumb of one measurement for every 3 ft of linear topography works very well for estimating purposes. Using this estimate, the typical field crew will make and record between 350–500 measurements or 1,000–1,500 ft of strip topography per day. Typically, a two-person crew equipped with recording total station and data collector pick up 1,250 ft a day. Depending on the office/field reduction software being used, these data can produce both the planimetric and contour maps as well as transfer the data to an engineering design package with very little additional manipulation.

To estimate strip topographic production remember the following tips:

- Estimate one measurement for every 3 ft of project.
- If shots are greater than 50 ft, a second rodperson adds to the efficiency of the crew.
- Expect a two-person crew equipped with a recording total station and data collector to pick up 1,250 ft per day.

Another question asked is, "Who runs the rod?" Conventional location or topographic surveying often requires a three-person field party. The party chief is working at the instrument, recording the measurements and other information in the field book. The party chief is also responsible for gathering data and must pay close attention to the movements of the rodperson. When the field crew consists of two people, it is often the party chief who runs the rod. When using the power of field-to-finish data collection, the experience and judgment of the rodperson is an important factor. Most organizations have the party chief or senior field technician run the rod and allow the less-experienced person to operate the instrument. The rodperson communicates codes and other instructions to the instrument operator, who enters them into the instrument or data collector and takes the measurements. Given the ease of operation of total stations, that is a fast and easy way to train an instrument operator. This frees more experienced personnel to control the pace of the job and to concentrate on gathering the correct data.

Multiple rodmen make for increased productivity. Data collection provides a tremendous increase in the speed of performing field work by eliminating the need to read and record measurements and other information. Many organizations have reduced the size of their field crews by eliminating the notetaker. However, this person can be very useful as a second rodperson.

On jobs where a large number of shots are needed, the use of two (or more) rodmen has resulted in excellent time and cost savings. The rodmen can work independently in taking ground shots or single features. They can work together by leapfrogging along planimetric or topographical feature lines. When more than one rodperson is used, crew members should switch jobs throughout the day. This helps to eliminate fatigue in the person operating the instrument.

As an extension of the concept discussed above it is a good idea to have an experienced person running one rod and directing the other rodmen. If possible, each rodperson and the instrument operator should have a radio or other means of reliable communication.

Electronic data collection (EDC) has proved to be an extremely cost-effective means of gathering data. When competing against a grid or baseline and offset style of surveying, EDC often results in field time savings of more than 50%. However, deriving a horizontal and vertical position on the located points is only part of the process. The ultimate goal is usually a map that shows planimetrics, contours, or both.

Total station productivity compared with other methods is shown in Figure 4-7, which depicts the enhanced productivity of a total station relative to traditional plane-table or transit-stadia methods. Time savings in design/construction layout is shown in Table 4-1.

## 4-24. Electronic Theodolite Error Sources

All theodolites measure angles with some degree of imperfection. These imperfections result from the fact that no mechanical device can be manufactured with zero error. In the past, very specific measuring techniques were taught and used by surveyors to compensate for minor mechanical imperfections in theodolites. With the advent of electronics, the mechanical errors still exist but are related to in a different way. One must now do more than memorize techniques that compensate for errors. One must clearly understand the concepts behind the techniques and the adjustments for errors that electronic theodolites now make. The following paragraphs provide the major sources of error when using a theodolite and also the particular method used to compensate for that error.

### a. Circle Eccentricity

Circle eccentricity exists when the theoretical center of the mechanical axis of the theodolite does not coincide exactly with the center of the measuring circle. The amount of error corresponds to the degree of eccentricity and the part of the circle being read. When represented graphically, circle eccentricity appears as a sine wave. Circle eccentricity in the horizontal circle always can be compensated for by measuring in both faces (opposite sides of the circle) and using the mean as a result. Vertical circle eccen-

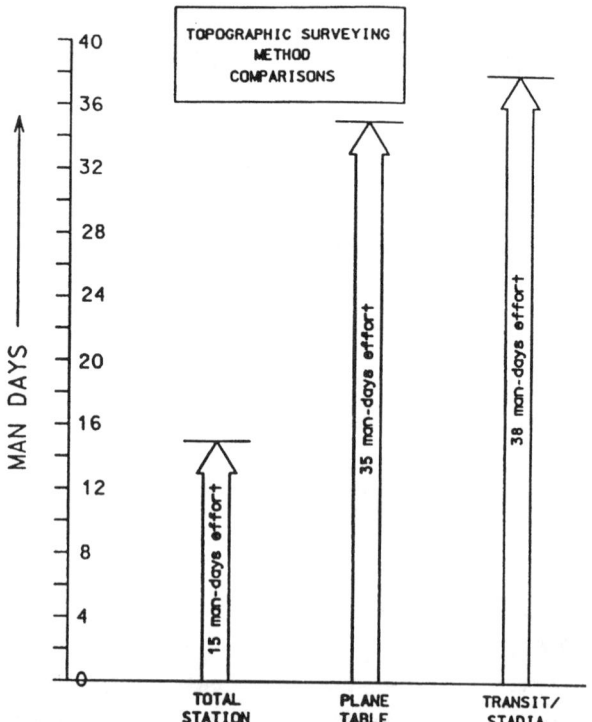

Figure 4-7. Topographic Surveying Method Comparisons

**Table 4-1. Total Stations Estimated Field Time Saved over Other Methods**

| Application on Project Site | Time in Field Saved (%) |
| --- | --- |
| Control Traverse | 5 |
| Topo | 25 |
| X-Select | 20 |
| Design Layout | 30 |
| As-Built | 35 |

*Note:* Topo = ground topographic.

tricity cannot be compensated for in this manner because the circle moves with the telescope. More sophisticated techniques are required.

Some theodolites are individually tested to determine the sine curve for the circle error in that particular instrument. Then, a correction factor is stored in ROM that adds or subtracts from each angle reading so that a corrected measurement is displayed.

Other instruments use an angle-measuring system that consists of rotating glass circles that make a complete revolution for every angle measurement. They are scanned by fixed and moving light sensors. The glass circles are divided into equally spaced intervals that are diametrically scanned by the sensors. The amount of time it takes to input a reading into the processor is equal to one interval, thus only every alternate graduation is scanned. As a result, measurements are made and averaged for each circle measurement. This eliminates scale graduation and circle eccentricity error.

### b. Horizontal Collimation Error

Horizontal collimation error exists when the optical axis of the theodolite is not exactly perpendicular to the telescope axis. To test for horizontal collimation error, point to a target in face one, then point back to the same target in face two; the difference in horizontal circle readings should be 180°. Horizontal collimation error can always be corrected for by meaning the face one and face two pointings of the instrument.

Most electronic theodolites have a method to provide a field adjustment for horizontal collimation error. Again, the manual for each instrument provides detailed instruction on the use of this correction.

In some instruments, the correction stored for horizontal collimation error can affect only measurements on one side of the circle at a time. Therefore, when the telescope is passed through zenith (the other side of the circle is being read), the horizontal circle reading will change by twice the collimation error. These instruments are functioning exactly as designed when this happens.

When prolonging a line with an electronic theodolite, the instrument operator should either turn a 180° angle or plunge the telescope and turn the horizontal tangent so that the horizontal circle reading is the same as it was before plunging the telescope.

### c. Height of Standards Error

In order for the telescope to plunge through a truly vertical plane, the telescope axis must be perpendicular to the standing axis. As stated earlier, there is no such thing as perfection in the physical world. All theodolites have a certain degree of error caused by imperfect positioning of the telescope axis. Generally, determination of this error should be accomplished by a qualified technician, because horizontal collimation and height of standards errors interrelate and can magnify or offset one another. Horizontal collimation error is usually eliminated before checking for height of standards. Height of standards error is checked by pointing to a scale the same zenith angle above a 90° zenith in face one and face two. The scales should read the same in face one as in face two.

### d. Circle Graduation Error

In the past, circle graduation error was considered a major problem. For precise measurements, surveyors advanced their circle on each successive set of angles so that circle graduation errors were "meaned out." Current technology eliminates the problem of graduation errors. This is accomplished by photo-etching the graduations onto the glass circles. Next, make a very large precise master circle, and photograph it. An emulsion is applied to the circle, and a photo-reduced image of the master is projected onto the circle. The emulsion is removed, and the glass circle has been etched with very precise graduations.

## 4-25. Total Survey System Error Sources and How To Avoid Them

In every survey, there is an accuracy that must be attained. The first step in using field and office time most effectively is to determine the positional tolerance of the points to be located. After this has been accomplished, all sources of error can be determined and analyzed. Some sources of error are pointing errors, prism offsets, adjustment of prism pole, EDM alignment, collimation of the telescope, optical plummet adjustment, instrument/EDM offsets, curvature and refraction, atmospheric conditions, effects of direct sunlight, wind, frozen ground, and vibrations. The accuracy required for each survey should be carefully evaluated. Each of the following factors can accumulate and degrade the accuracy of measurements.

### a. Pointing Errors
Pointing errors are due to both human ability to point the instrument and environmental conditions limiting clear vision of the observed target. The best way to minimize pointing errors is to repeat the observation several times and use the average as the result.

### b. Uneven Heating of the Instrument
Direct sunlight can heat one side of the instrument enough to cause small errors. For the highest accuracy, pick a shaded spot for the instrument.

### c. Vibrations
Avoid instrument locations that vibrate. Vibrations can cause the compensator to be unstable.

### d. Collimation Errors
When sighting points a single time (e.g., direct position only) for elevations, check the instrument regularly for collimation errors.

### e. Vertical Angles and Elevations
When using total stations to measure precise elevations, the adjustment of the electronic tilt sensor and the reticle of the telescope becomes very important. An easy way to check the adjustment of these components is to set a baseline. A line close to the office with a large difference in elevation will provide the best results. The baseline should be as long as the longest distance that will be measured to determine elevations with intermediate points at 100- to 200-ft intervals. Precise elevations of the points along the baseline should be measured by differential leveling. Set up the total station at one end of the baseline and measure the elevation of each point. Comparing the two sets of elevations provides a check on the accuracy and adjustment of the instrument. Accuracy requirements may dictate that more than one set of angles and distances be measured to each point. Some examples are distances >600 ft, adverse weather conditions, and steep observations.

### f. Atmospheric Corrections
Instruments used to measure atmospheric temperature and pressure must be correctly calibrated.

### g. Optical Plummet Errors
The optical plummet or tribrachs must be periodically checked for misalignment.

### h. Adjustment of Prism Poles
When using prism poles, precautions should be taken to ensure accurate measurements. A common problem encountered when using prism poles is the adjustment of the leveling bubble. Bubbles can be examined by establishing a check station under a doorway in the office. First, mark a point on the top of the doorway. Using a plumb bob, establish a point under the point on the doorway. If possible, use a center punch to make a dent or hole in both the upper and lower marks. The prism pole can now be placed into the check station and easily adjusted.

### i. Recording Errors
The two most common errors associated with field work are reading an angle incorrectly and/or entering incorrect information into the field book. Another common (and potentially disastrous) error is an incorrect rod height. Although electronic data collection has all but eliminated these errors, it is still possible for the surveyor to identify an object incorrectly, make a shot to the wrong spot, or input a bad target height or HI. For example, if the surveyor normally shoots a fire hydrant at the ground level but for some reason shoots it on top of the operating nut, erroneous contours would result if the program recognized the fire hydrant as a ground shot and was not notified of this change in field procedure.

### j. Angles
As a rule, a surveyor will turn a doubled angle for move-ahead, traverse points, property corners, or other objects that require greater accuracy. On the other hand, single angles are all that are required for topographic shots.

## 4-26. Controlling Errors

A set routine should be established for a survey crew to follow. Standard operating procedure should require that control be measured and noted immediately on the data collector and in the field book after the instrument has been set up and leveled. This ensures that the observations to controlling points are established before any outside influences have had an opportunity to degrade the setup. In making observations for an extended period of time at a particular instrument location, observe the control points from time to time. This ensures that any data observed between the control shots are good or shows that a problem has developed, and appropriate action can be taken to remedy the situation. As a minimum, require survey crews to observe both vertical and horizontal controls points at the beginning of each instrument setup and again before the instrument is picked up.

One of the major advantages of using a total station equipped with data collection is that errors previously attributed to blunders (i.e., transposition errors) can be eliminated. Even if the wrong reading is set on the horizontal circle in the field or the wrong elevation is used for the bench, the data itself may be precise. To make the data accurate, many software packages will allow the data to be rotated and/or adjusted as it is processed. The only way to assure that these corrections and/or observations have been accurately processed is to compare the data with control points. Without these observations in the magnetically recorded data, the orientation of that data always will be in question.

The use of a total station with a data collector can be looked upon as two separate and distinct operations. The checklists for setting up the total station and data collector are as follows:

1. Total Station

    a. If EDM is modular, mount it on instrument.
    b. Connect data collector.
    c. Set up and level instrument.
    d. Turn on total station.
    e. Set atmospheric correction (ppm). This should be done in the morning and at noon.
    f. Set horizontal circle.
    g. Set coordinates.
    h. Observe backsight (check whether azimuth to backsight is 180° from previous reading).
    i. Observe backsight benchmark (obtain difference in elevation). This may require factoring in the height of reflector above benchmark.
    j. Compute relative instrument height (benchmark elevation ± difference in height). Note height of rod and note computations in field book.
    k. Input Z (elevation) value in instrument or data collector.
    l. Observe backsight benchmark (check elevation).
    m. Invert and repeat (check elevation).

2. Data Collector

    a. Record date and job number.
    b. Record crew number and instrument serial number.
    c. Record field book number and page number.
    d. Record instrument location (coordinates).
    e. Record backsight azimuth.
    f. Record standard rod height.
    g. Record height of instrument.

Note: All the above information (steps a–g) also should be recorded in the field book.

    h. Observe and record measurement to backsight benchmark.
    i. Enter alpha or numeric descriptor of above point into data collector.
    j. Observe and record measurement backsight benchmark or check benchmark. (If setting benchmark, note in field book and repeat with instrument inverted.)
    k. Enter alpha or numeric descriptor of above point into data collector.
    l. Observe and record measurement to backsight.
    m. Enter alpha or numeric descriptor of above point into data collector.
    n. Invert and repeat steps l and m.
    o. Observe and record measurement to foresight.
    p. Enter alpha or numeric descriptor of above point into data collector.
    q. Invert and repeat steps o and p.
    r. Observe and record measurement to side shot.
    s. Enter alpha or numeric descriptor of above point into data collector (repeat steps r and s as needed).
    t. When setup is complete, or at any appropriate time, repeat shots on vertical and horizontal control. Observe the displays and record in data collector.

## 4-27. Coding Field Data

Whether data are recorded by hand or electronically, one of the most time-consuming survey operations is the recording of a code or description to properly identify the point during processing. For example, in a topographic or planimetric survey, the identification of points that locate the position of curbs, gutters, center lines, manholes, and other similar features is essential for their correct plotting and contour interpolation.

Despite this slow coding process encountered when using today's data collectors, the advantages heavily outweigh the disadvantages. The advantages include the collection of error-free numeric data from electronic total stations virtually at the instant they are available, and the error-free transfer of these data to an office computer system without the need for manual entry.

Field coding allows the crew to perform the drafting and provides a more logical approach. Because the field crew can virtually produce the map from the field data, this eliminates the need for many field book sketches. They can also eliminate office plotting, editing by connecting the dots, etc., to produce a final product. Total station users who gather data to be processed with other systems typically record descriptive information with each point measured and gather 200–300 points per day with the total station. Users report 300–700 points per day if descriptive information is kept to the necessary minimum. The coding scheme is designed so the computer can interpret the recorded data without ambiguity to create a virtually finished product.

## 4-28. Field Computers

Many districts perform much of the survey reduction in the field. The greatest advantage of this procedure is uncovering a mistake that can easily be corrected if the crew and equipment are on the site. Laptop and notebook computers are popular field items. These computers are used to download GPS and total station data. After the files are stored in the computer, data reduction can be done easily with programs stored in these machines.

Listed below are some software considerations to install on topographic field computers:

- Interface with field data collector
- A system of predefined codes for most common objects and operations in a database
- User-defined codes for site-specific requirements in a database
- Survey adjustment programs, such as
    - compass rule adjustment
    - transit rule adjustment
    - Crandell method
    - least squares
    - angle adjustment
    - distance adjustment
- A program that can assign an alphanumeric descriptor field for each survey point
- A full-screen editor to examine and edit ASCII data files
- An interface program to convert files to common graphic interchange formats such as IGES or DXF
- A program to connect features that were not recorded in order, such as fence, curb and gutter, edge of pavement, waterline etc.
- An operating system that will be compatible with post-processing machines with CADD programs such as Intergraph, Accugraph, and AutoCAD
- Custom programs that can use all the features available to the total station or the data collector
- Software that provides training, if possible

Requirements for field computers in data collection are as follows:

1. Portable
2. Rugged, for field use
3. Processor: 80386 or 80486; must run MS-DOS 4.0 and greater
4. Memory: 640-kilobyte main memory (minimum), 40-megabyte internal hard disk (minimum; 80 megabytes recommended)
5. Disk drives: 3.5-in. floppy 720 k or 1.44 meg (1.44 preferred) external 5.25-in. floppy
6. Math coprocessor needed (must be compatible with main processor) (386)
7. Serial port (RS 232), parallel port
8. Modem: 2,400-baud autodial (Hayes compatible)
9. VGA or Super VGA graphics
10. Portable inkjet printer

## 4-29. Modem for Data Transfer (Field to Office)

A modem is a device that modulates and demodulates binary data transmission over a telephone network. This device MODulates the carrier for transmission and DEModulates for reception, hence the term MODEM. The carrier is simply a tone with three characteristics, any one of which can be varied or modulated to impose a signal on the carrier. They include amplitude, frequency, and phase. The exact method of variation must be the same for two or more modems to be called "compatible."

## 4-30. Trigonometric Leveling and Vertical Traversing

Trigonometric leveling is the single most important new application brought into widespread use by the increasing acceptance of the total station. It is a fair statement that the error sources and types that affect trigonometric leveling measurements are among the least understood of the commonly done surveying procedures (see Table 4-2). A knowledge of the limitations of trigonometric leveling, together with means

# TOPOGRAPHIC SURVEY TECHNIQUES

(instrumentation and procedures) to account for such limitations, is essential in using and supporting the use of modern surveying technology.

Total station trigonometric leveling can achieve accuracies similar to those reached using a spirit level. Third-order accuracy should be easily obtainable. First-order accuracy has been achieved, but the procedures are involved and not commonly followed.

Figures 4-8 and 4-9 depict the advantage of trigonometric leveling over that of spirit levels, especially high-relief terrain.

## 4-31. Trigonometric Leveling Field Procedures

To obtain third- or second-order vertical accuracies with a total station, the following field procedures should be rigorously followed:

- Careful setup and leveling
- Use face I and face II observations
- Reciprocal measurements
- Take multiple observations
- Protect instrument from sun and wind
- Use proper targetry based on Inst/EDM configuration
    - Tilting target if necessary
    - Good-quality reflectors
    - Correct prism offsets
    - Unambiguous target
    - Maintain targetry in good adjustment
- Limited sight distances
    - 300 m maximum
    - Reduce atmospheric-related error
    - Improves vertical angle accuracy

To measure the difference in height between two points (A and B) using spirit leveling, do this:

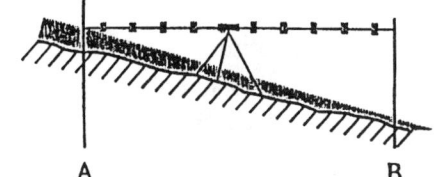

This process would be continued over many setups to form a line of levels, or a level run, where the points A and B are not visible from the one level setup.

**Figure 4-8. Spirit Leveling**

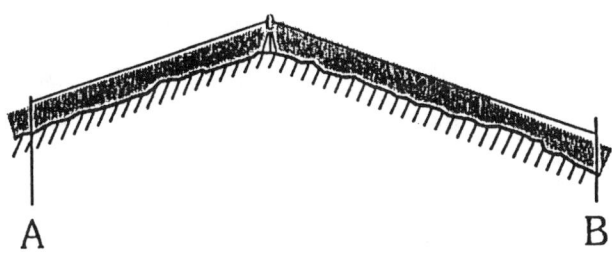

Will accomplish the same results with considerable savings

**Figure 4-9. Trig Leveling**

- Accurately measure temperature and pressure
    - At least twice a day
    - If long steep line measurements at both ends, use averages.
- Watch for adverse refraction

## 4-32. Trigonometric Leveling Error Sources

The following error sources impact the accuracy of trigonometric leveling with electronic total stations:

- Instrument
    - Distances
    - Vertical angle accuracy
    - Vertical compensator important, dual axis compensation

**Table 4-2. Elevation Errors Due to Errors in Zenith Angles**

| Sight Distance (ft) | Vertical Angle Uncertainty | | | |
|---|---|---|---|---|
| | 1 sec | 5 sec | 10 sec | 60 sec |
| 100 | 0.0005 | 0.0024 | 0.005 | 0.03 |
| 200 | 0.0010 | 0.0048 | 0.010 | 0.06 |
| 400 | 0.0019 | 0.0097 | 0.019 | 0.12 |
| 500 | 0.0024 | 0.0121 | 0.024 | 0.15 |
| 1,000 | 0.0048 | 0.0242 | 0.049 | 0.29 |

- No boost to vertical angle accuracy
- Nature
  - Curvature and refraction
  - Temperature/pressure correction
  - Wind, sun, and weather
- Operator
  - Accurate pointing

- Lots of measurements
- Reciprocal measurements
- Measure HI and HT

Table 4-3 shows the precision resulting from horizontal distance and vertical angular measurements as needed to resolve differences in elevations from trigonometric observation.

### Table 4-3. Combining Sources of Error (500-ft line)

| Source | Type | Nominal Amount |
|---|---|---|
| **Errors in Vertical Angle Measurement**[a] | | |
| Instrument Accuracy | Random | ± 3 sec |
| Collimation | Systematic | ± 3 sec |
| Measure HI and HT | Random | ± 0.005 to 0.1 ft |
| C and R | Systematic | 0.005 ft |
| Hand-Held Prism Pole | Random | ± 0.005 ft |
| 30-mm Prism Offs Error | Random | ± 3 mm |
| Heatwaves | Random | ± 00.01 ft |
| Unshaded Instrument | Random | ± 5 in. to 10 sec |
| **Errors in Distance Measurement**[b] | | |
| Nominal Accuracy | Random | ± (5 mm + 5 ppm) |
| Temp Estimation | Random | ± 10 degrees F |
| Pressure Estimation | Random | ± 0.5 in Hg |
| Prism and Instrument Calibration | Systematic | ± 2 mm |
| Prism Mispointing | Random | ± 0.35 mm |
| Hand-Held Prism Pole | Random | ± 5 ft (fore/ft lean) |

[a] Combined angular and linear error is between –0.024 and 0.048 ft. Vertical angle precision for the 500-ft line is therefore between 1:10,000 and 1:210,000.

[b] Combined error is between –0.0414 and +0.0546 ft. On this 500-ft line, this gives a range in precision of 1:9,000 to 1:12,000.

# CHAPTER 5

# DATA COLLECTION PROCEDURES FOR THE TOTAL STATION

## 5-1. General

In the first step of the process, the field survey, the vertical and horizontal angles are measured along with slope distances using the total station. The angle and distances are stored with a point number and description in the data collector. The survey data are then transferred to the microcomputer via a cable connection for data processing and field data storage. The microcomputer is either an in-office desktop system or a laptop model that can be used on site.

The data are then processed in the microcomputer to produce a coordinate file that contains point number, point code, $X, Y, Z$ coordinate values, and a point descriptor.

After the data are on the workstation, they are converted into a graphics design file for use in a computer-aided drafting and design (CADD) program such as MicroStation or AutoCad. The program CVTPC, available through the U.S. Army Topographic Engineering Center, can be used to convert the ASCII files into Intergraph design files. Level, label, symbol, and line definitions are assigned to each point based upon point code. The program can transform data into a two-dimensional (2D) or three-dimensional (3D) design file.

The 3D file is used to create the digital terrain model (DTM), which is used to produce the contours. The resulting topographic data are then plotted for review. Final editing and addition of notes are completed, yielding topographic data in a digital format or as a plotted map.

Uniform operating procedures are needed to avoid confusion when collecting survey data. The use of proper field procedures is essential to prevent confusion in generating a map. Collection of survey points in a meaningful pattern aids in identifying map features.

## 5-2. Functional Requirements of a Generic Data Collector

The question of field notes is an important issue. Some districts require field notes to be kept, whereas others use a data collector to replace field notes. An important distinction is made *if field notes are not required*, and the data collector is used as an "electronic field book." Total stations calculate coordinates in situ and can continuously store coordinates, either in their own memory or in a data collector. *If field notes are required*, only specific items are considered in the transfer of data from a data collector to an office computer. The advantage of this method is that a check is provided on field notes. Most field note errors are made by transcription, for example, writing 12 instead of 21 in the field book. Data transmitted to an office computer, through an RS-232C port, can be listed on an office printer to provide a check for transposition errors in the field notes. If the data collection is bidirectional, then it must receive data from the office computer for stake-out purposes as well as transmit data to the computer. Field notes can again be considered or ignored.

Some districts feel that the electronically collected data is the field book required. Other districts still require that a field book be kept for data safeguarding and legal issues. When field books are kept, the entries are compared with the files generated by the data collection processing. All data are booked in the format of the standards set by each district or branch. In many cases, there are rarely two districts that have a standard format of notekeeping that is identical.

Four kinds of notes are kept in practice: sketches, tabulations, descriptions, and combinations of these. The most common kind is a combination form, but an experienced recorder selects the version best suited to the job at hand. The location of a reference point may be difficult to identify without a sketch, but often a few lines of description are enough. Benchmarks are also described. In notekeeping, this axiom is always pertinent: When in doubt about the need for any information, include it and make a sketch. It is better to have too much data than not enough.

Observing the suggestions listed here will eliminate some common mistakes in recording notes:

- Letter the notebook owner's name and address on the cover and first inside page, in India ink.
- Use a hard pencil or pen, legible and dark enough to copy.
- Begin a new day's work on a new page.
- Immediately after a measurement, always record it directly in the field book rather than on a sheet of scrap paper for copying it.
- Do not erase recorded data.
- Use sketches instead of tabulations when in doubt.
- Avoid crowding.
- Title, index, and cross-reference each new job or continuation of a previous one.
- Sign surname and initials in the lower right-hand corner of the right page on all original notes.

Figures 5-1 and 5-2 are examples of sketches in a field book used for digital surveys. Topographic locations are numbered according to data record numbers. Data record numbers (point numbers) depict what kind of location and where locations were measured. This helps office personnel improve digital field drawings into final design drawings. More important, blunders and mislabeled feature codes may be caught before costly design errors are made. The finished map and the sketch should be similar. Sketches are not required to be at any scale.

Electronic files are sufficient for submittal without identical hand entries from a field book. Video and digital cameras can be used to supplement the field sketch and provide a very good record of the site conditions for the CADD operator, design engineer, and user of the topographic map.

## 5-3. Data Collection Operating Procedures

Uniform operating procedures are needed to avoid confusion when collecting survey data. The use of

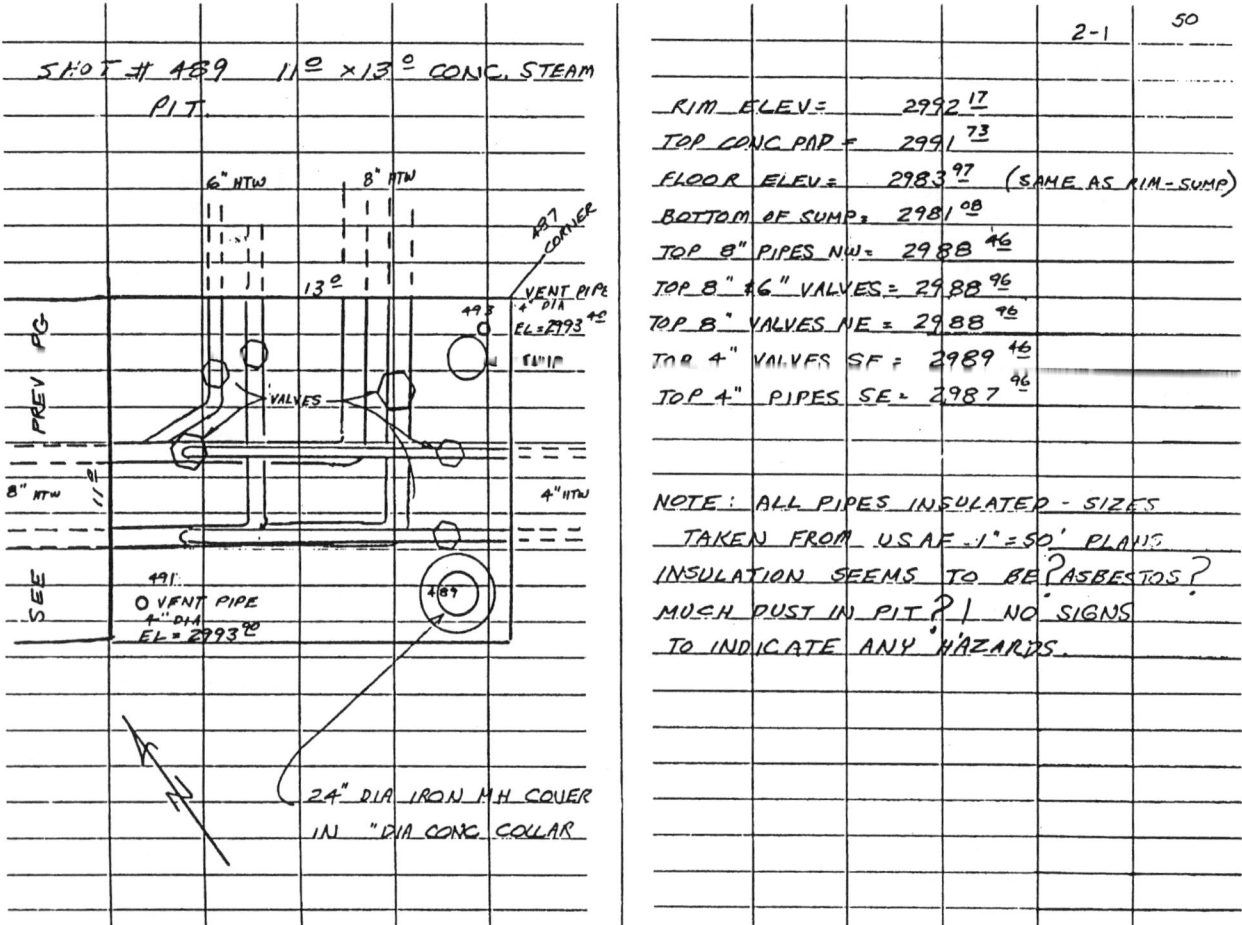

**Figure 5-1. Sample of Field Book Sketch Used for Digital Surveys, Example 1**

# DATA COLLECTION PROCEDURES FOR THE TOTAL STATION

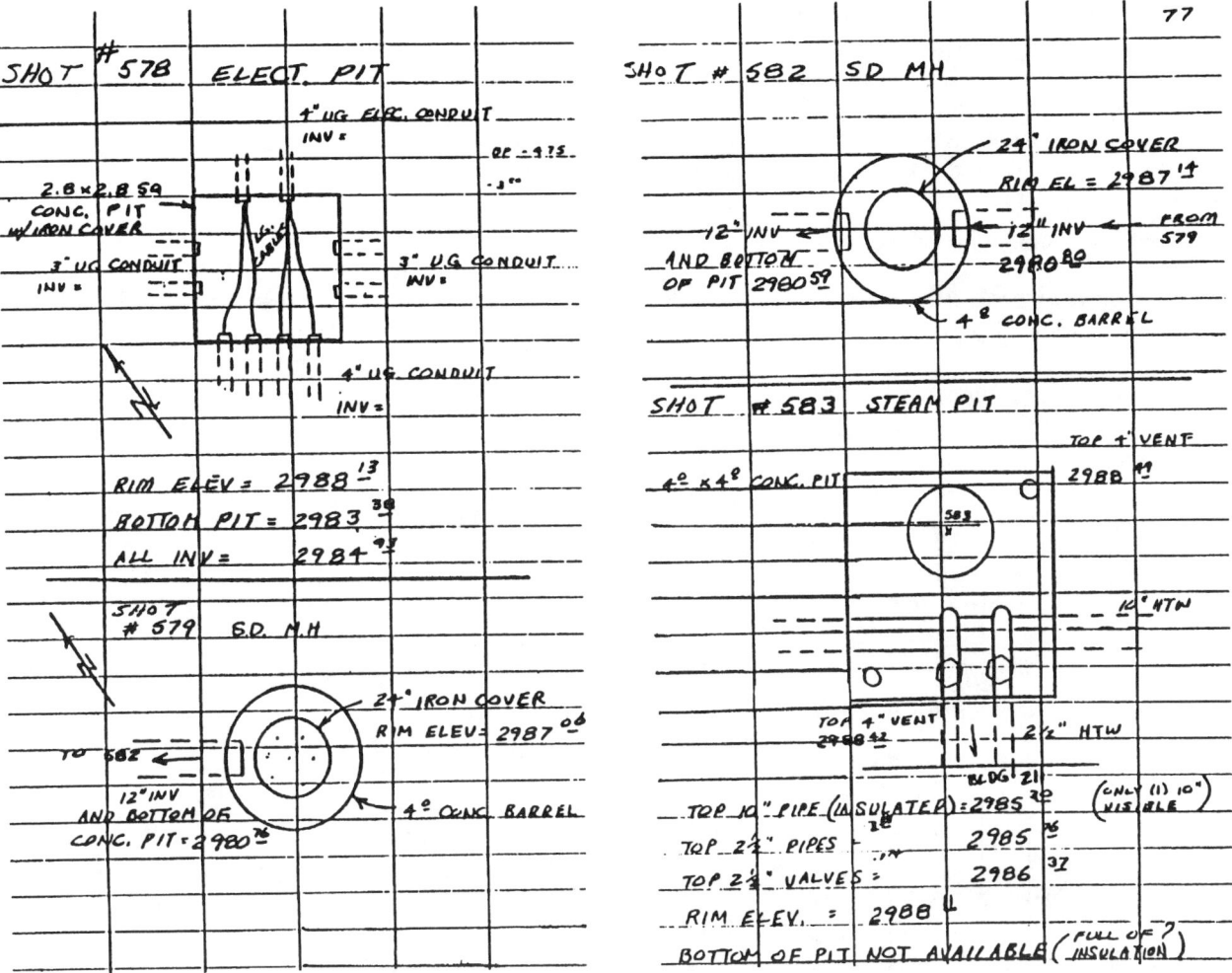

Figure 5-2. Sample of Field Book Sketch Used for Digital Surveys, Example 2

proper field procedures is essential to prevent confusion in generating a map. Collection of survey points in a meaningful pattern aids in identifying map features. Experience has resulted in the following steps for collection of field data.

### a. Establish Horizontal and Vertical Control for Radial Survey

This includes bringing control into the site and establishing setup points for the radial survey. Primary control is often brought into the site using the GPS satellite receivers. The traverse through radial setup points can be conducted with a total station as the radial survey is being performed. Experience indicates a separate traverse is preferable. A separate traverse results in less opportunity for confusion of point identification and allows the quality of the traverse to be evaluated before it is used. Elevations are established for the radial traverse points by using conventional leveling techniques instead of the trigonometric values determined from the total station.

### b. Perform Radial Surveys to Obtain Information for Mapping

1. Set the total station over control points established as described earlier.

2. Measure and record the distance from the control point up to the electronic center of the instrument, as well as the height of the prism on the prism pole.

3. Maintain accuracy. To prevent significant errors in the map elevations, the surveyor must report and record any change in the height of the prism pole. For accuracy, use a suitable prism and target that matches optical and electrical offsets of the total station.

### c. Collect Data in a Specific Sequence

1. Collect planimetric features (roads, buildings, etc.) first.

2. Enter any additional data points needed to define the topography.

3. Define break lines. Use the break lines in the process of interpolating the contours to establish regions for each interpolation set. Contour interpolation will not cross break lines. Assume that features such as road edges or streams are break lines. They do not need to be redefined.

4. Enter any additional definition of ridges, vertical, fault lines, and other features.

**d. Draw a Sketch of Planimetric Features**

A sketch or video of planimetric features is an essential ingredient to proper deciphering of field data. The sketch does not need to be drawn to scale and may be crude but must be complete. A crude sketch is shown in Figure 5-3. The sketch is of an office courtyard. Numbers listed on the sketch show point locations. The sketch helps the CADD operator who has probably never been to the job site confirm that the feature codes are correct by checking the sketch.

A detail sketch is shown in Figure 5-4. This information is critical to the design engineer. Detail sketches can be used to communicate complex information directly to the engineer without lengthy discussions.

Miscellaneous descriptive notes can also be shown on the sketch for later addition to the design

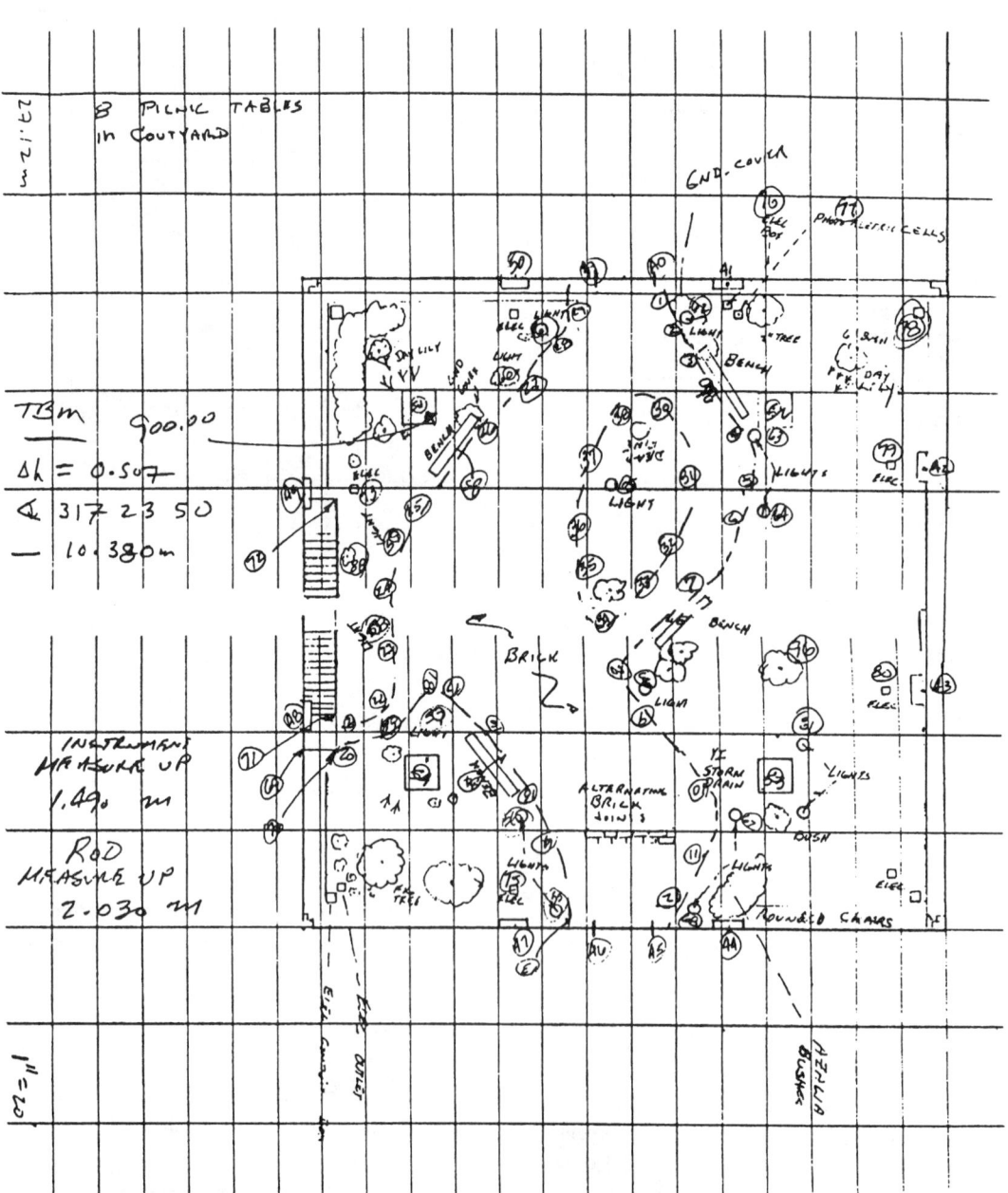

**Figure 5-3. Sketch of an Office Courtyard with Point Numbers**

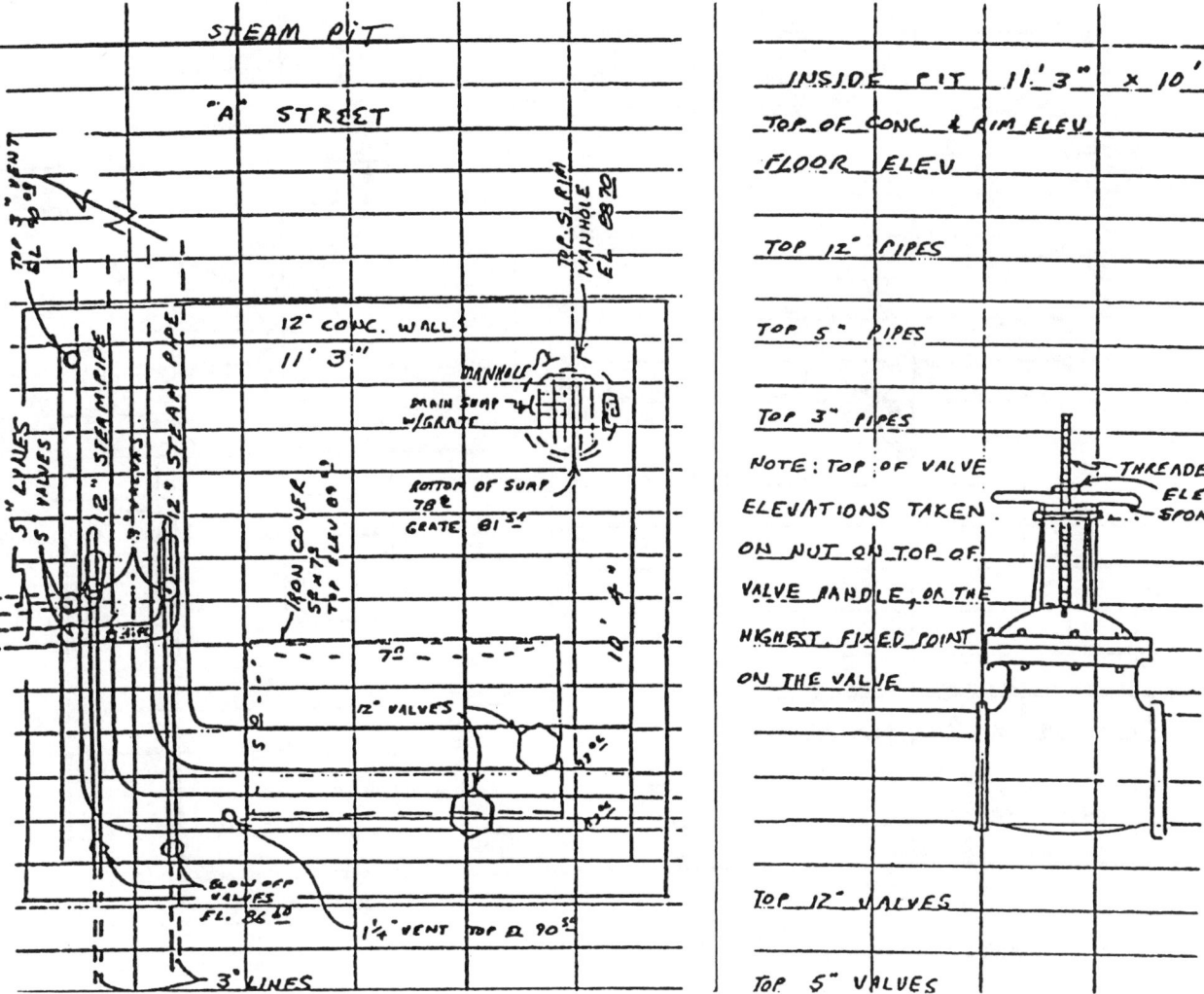

Figure 5-4. Detail Sketch

file. These notes are usually clearer and contain more information when shown on the sketch than when entered into the data collector. Figures 5-5 and 5-6 indicate more information than can be typed into a data collector at the present time.

### e. Obtain Points in Sequence

The translation to CADD program will connect points that have codes associated with linear features (such as the edge of a road) if the points are obtained in sequence. For example, the surveyor should define an edge of a road by giving shots at intervals on one setup. Another point code, such as natural ground, will break the sequence and will stop formation of a line on the subsequent CADD file. The surveyor should then obtain the opposite road edge.

### f. Use Proper Collection Techniques

Using proper techniques to collect planimetric features can give automatic definition of many of these features in the CADD design file. This basic picture helps in operation orientation and results in easier completion of the features on the map. Improper techniques can create problems for office personnel during analysis of the collected data. The function performed by the surveyor in determining which points to obtain and the order in which they are gathered is crucial. This task is often done by the party chief. Cross-training in office procedures gives field personnel a better understanding of proper field techniques.

Most crews will make and record 250–400 measurements per day. This includes any notes that must be put into the system to define what was measured. A learning curve is involved in the establishment of productivity standards. It usually takes a crew five to six projects to become confident enough with their equipment and the coding system to start reaching system potential.

A two-person crew is most efficient when the typical spacing of the measurements is less than 50 ft.

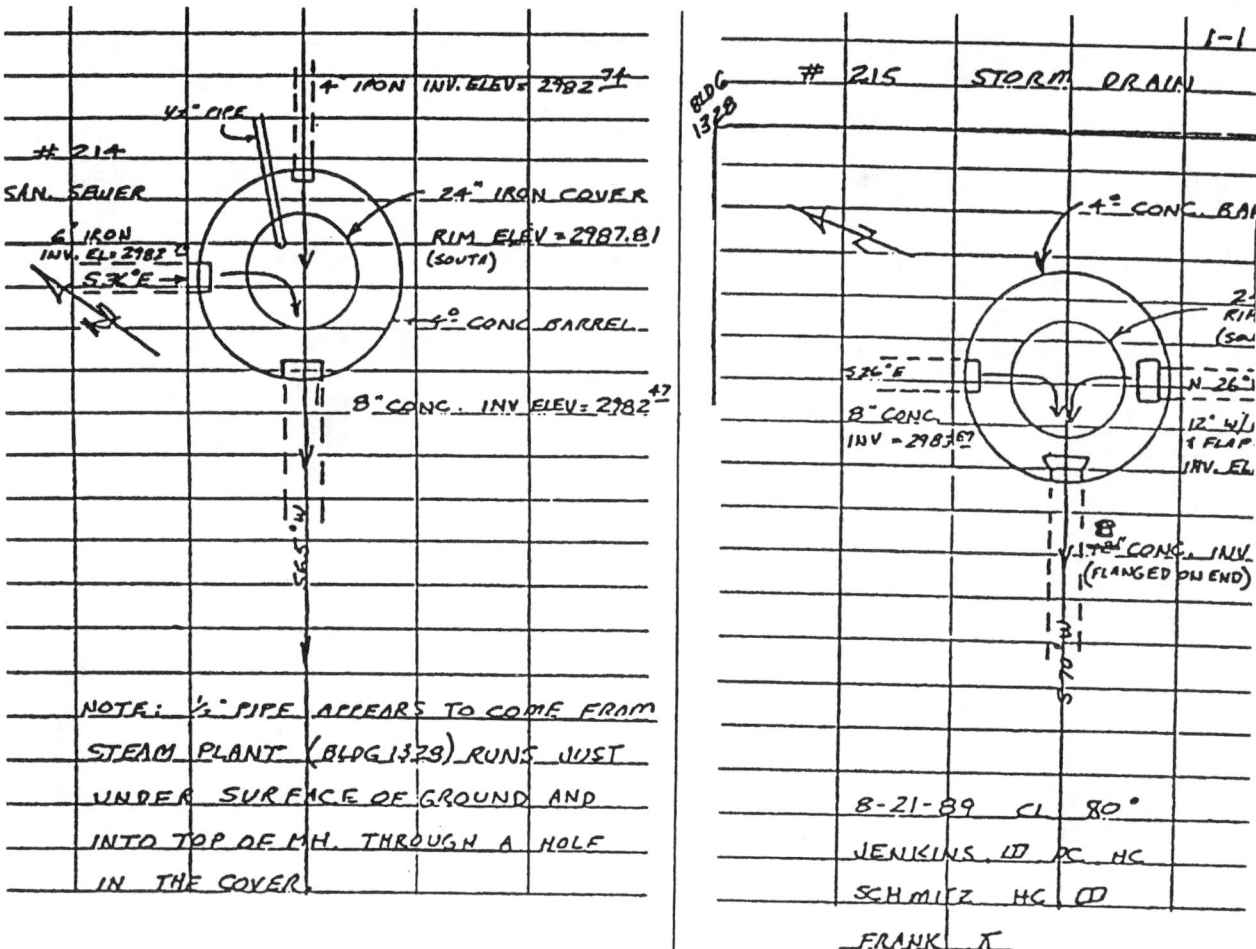

Figure 5-5. Field Notes to Accompany Data Collector

When working within this distance, the average rodperson can acquire the next target during the time it takes the instrument operator to complete the measurement and input the codes to the data collector. The instrument operator usually spends about 20 seconds sighting a target and recording a measurement and another 5–10 seconds coding the measurement.

When the general spacing of the measurements exceeds 50 ft, having a second rodperson will increase productivity. A second rodperson allows the crew to have a target available for measurement when the instrument operator is ready to start another measurement coding sequence. After the measurement is completed, the rodperson can move to the next shot, and the instrument operator can code the measurement while the rodpeople are moving. If the distance of that move is 50 ft or greater, the instrument will be idle if you have only one rodperson.

Data collection provides a tremendous increase in the speed of field work by eliminating the need to read and record measurements and other information.

On jobs where a large number of shots are needed, the use of two (or more) rodpeople has resulted in excellent time and cost savings. Communication between the rodperson and the instrument person is commonly done via T/R radio. The rodpeople can work independently in taking ground shots or single features or they can work together by leapfrogging along planimetric or topographic feature lines. When more than one rodperson is used, crew members should switch jobs throughout the day. This helps to eliminate fatigue in the person operating the instrument.

## 5-4. Field Crew Responsibility

Upon the completion of the file transfer, make a backup copy of the raw data. After this transfer is complete, and ONLY AFTER this transfer is complete, then the data in the data collector can be deleted.

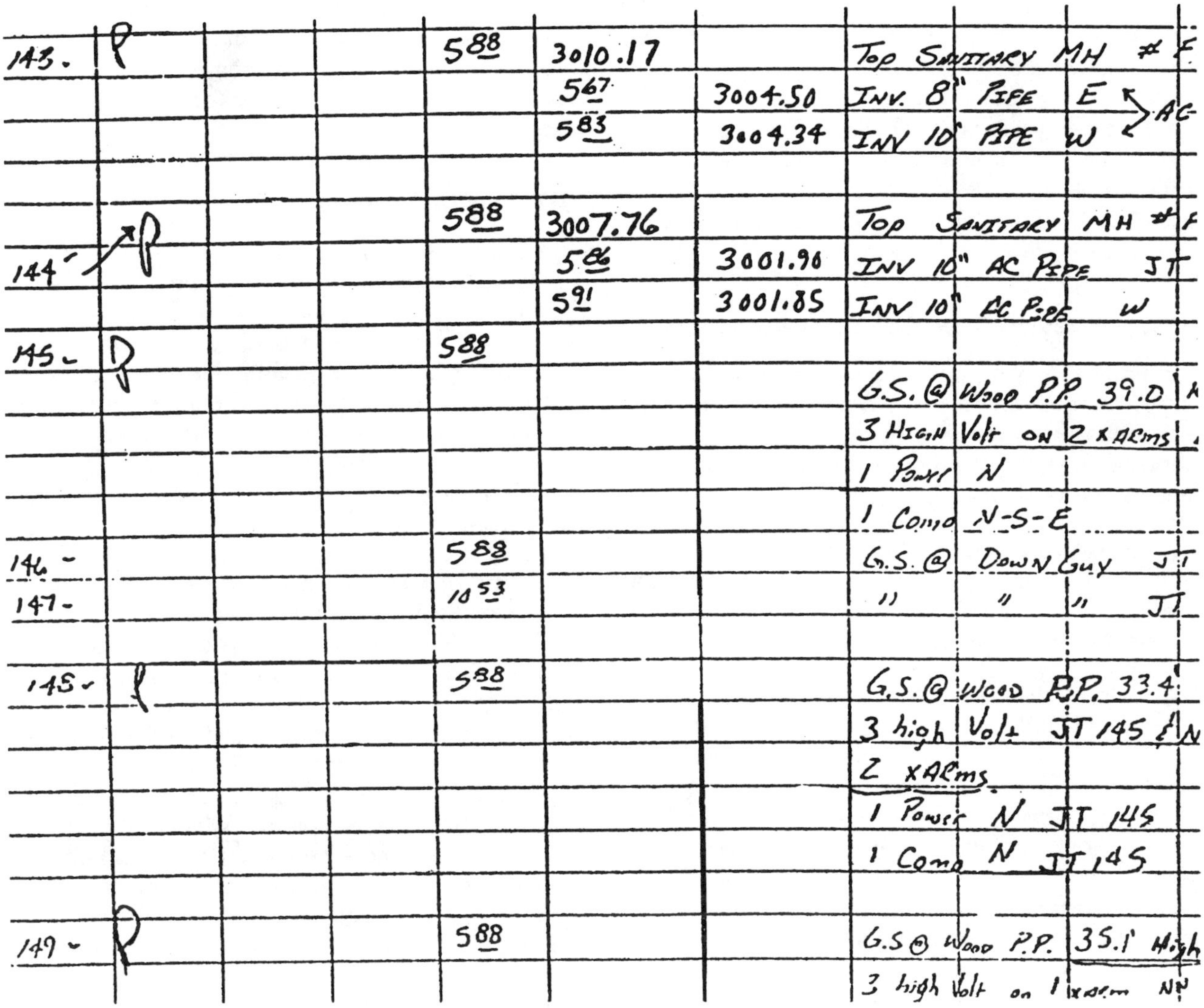

| 143. | ዋ | | | | 5 88 | 3010.17 | | Top Sanitary MH # F |
|---|---|---|---|---|---|---|---|---|
| | | | | | 5 67 | 3004.50 | | Inv. 8" Pipe E |
| | | | | | 5 83 | 3004.34 | | Inv 10" Pipe W →AC |
| | | | | | 5 88 | 3007.76 | | Top Sanitary MH # F |
| 144 | ዋ | | | | 5 86 | 3001.90 | | Inv 10" AC Pipe JT |
| | | | | | 5 91 | 3001.85 | | Inv 10" AC Pipe W |
| 145 - | ዋ | | | | 5 88 | | | G.S. @ Wood P.P. 39.0' |
| | | | | | | | | 3 High Volt on 2 xArms |
| | | | | | | | | 1 Power N |
| | | | | | | | | 1 Comm N-S-E |
| 146 - | | | | | 5 88 | | | G.S. @ Down Guy JT |
| 147 - | | | | | 10 53 | | | " " " JT |
| 148 - | ዋ | | | | 5 88 | | | G.S. @ Wood P.P. 33.4' |
| | | | | | | | | 3 high Volt JT 145 N |
| | | | | | | | | 2 xArms |
| | | | | | | | | 1 Power N JT 145 |
| | | | | | | | | 1 Comm N JT 145 |
| 149 - | ዋ | | | | 5 88 | | | G.S @ Wood P.P. 35.1' High |
| | | | | | | | | 3 high Volt on 1 xArm N |

**Figure 5-6. Field Notes for Point Locations**

Print a copy of the formatted data and check it against the field notes. Check the field input of data against the field notes. Specifically, check the instrument locations, azimuths to backsights, and the elevation of benchmarks. Also scan the data for any information that seems to be out of order. Check rod heights.

Edit the data. Eliminate any information that was flagged in the field as being in error. In the system, make a record of any edits, insertions, deletions, who made them, and when they were made.

Process the control data. Produce a short report of the data that were collected in the field. Check the benchmark elevation to be certain that the given elevation is the calculated elevation and that the coordinates of the backsights and foresights are correct.

To assure that good data are being supplied by the field, make certain that the field crew fully understands the automated processes that are being used and that they take care to gather data appropriately. It is much easier and more productive for the field crew to get a few extra shots where they know there will be difficulty in generating a good contour map than it will be for those in the office to determine where certain shots should have been made and add them to the database. Also make sure they pick up all breaklines necessary to produce the final map.

The field crew will need to become educated about the contouring package used by the branch. As the data are brought in from the first few projects and periodically thereafter, the crew should observe the product produced by the contouring program. This will help them to understand where and what amount of data may be needed to get the best results.

The office staff needs to be aware that in some circumstances the field will have difficulty in getting some information (terrain restrictions, traffic, etc.).

The person responsible for the field work should be involved in the initial phase of editing, because he or she will most likely remember what took place. Preferably, the editing should be done the same day the data are gathered, while the field person's memory is still fresh. If it is not possible for someone to walk the site to ensure that the final map matches the actual conditions, then the field person should be the one to review the map.

### 5-5. Surveyor–Data Collector Interface

For many surveying operations, electronic data collection is routine. However, after the data are collected, most software systems require a large amount of post-processing to produce a map showing planimetrics and contours.

#### a. Computer Interfacing

Many of the benefits of automated data collection are lost if the data stored cannot be automatically transferred to a computer system.

#### b. Hardware Compatibility

Most micro- and minicomputers on the market today are supplied with or have as an option a serial interface board. The serial interface typically supports communications at different baud rates (speeds of transmission) and with different parity settings. To control the flow of data, either a hardware or software handshake is used. Cables are connected to the serial interface board using a standard 25-pin connector. Occasionally, nonstandard connectors with a different number of pins are used. Every data collector stores data in a different format, and the problem is to translate the data from the data collector format into a file with a standard ASCII format. Data standardization will become more important in the future, and surveyors should be searching for methods that make system integration easier.

### 5-6. Digital Data

The fact that survey data collected by computer is in digital form has until recently been of interest only to surveyors themselves. Because the final products delivered to the clients were drawings, surveyors have needed to invest in only the computer equipment and software they needed to get the digital data collected and plotted as a scaled drawing. Now, the situation is changing. The proliferation of computer graphics used by architects, engineers, and developers has meant that surveyors are asked, even required, to deliver survey information in digital format. These demands can pose thorny technical problems for those who did not consider this eventuality when they acquired their computer systems. The time and expense to work out the technical details of digital data delivery can be prohibitive to those who consider themselves as surveyors, not computer experts.

### 5-7. Digital Transfer

There are two ways to transfer survey information digitally: as numeric data or as graphic files. The first is simpler from the surveyor's point of view. It begins with a text file—the sort of data that can be produced using a word processor. Text files are easiest to transfer between computers, but the clients want data that computer software can interpret to produce drawings, not raw field notes.

Again, if the surveying software permits the output of the appropriate information in a text file, reformatting that information is, at worst, a minor programming task and may be possible simply through the global replacement feature of a word processing package. However, surveyors who use word processors to edit text files should be sure to use the "ASCII" output option that is available in most word processors. This creates a "generic" text file without embedded control or formatting characters.

Most CADD systems require digital deliverables and graphics files compatible with their particular system. This is a more problematic request because, unlike COCO tables, which are uniform no matter which vendor's version of COCO is being used, every CADD system has a unique and proprietary graphic data format. This means, for example, that graphic data produced in AutoCad cannot be loaded onto an Intergraph system without some sort of intermediary "translation." Thus, even when data collectors are interfaced with a major CADD package, the diversity of CADD systems being used in the United States today virtually guarantees there will be clients using different systems and unable to load the graphic file directly.

Translation of graphics data can be handled in two ways: by direct translation or through a neutral format. A direct translator is a computer program that reads graphic data in one specific CADD system's format and outputs the same graphic information in a second CADD system's format. Although this is generally the quickest and most foolproof way to perform translation, it is often the most expensive. Because direct translation programs address only the problem of translation between two specific systems, several different translation programs may be necessary to provide data that meet the compatibility requirements of all the surveyor's clients.

Because all CADD vendors regard their data formats as proprietary, this process generally requires programmers who are intimately familiar with both CADD systems to write translation software.

It may be hard to locate all the programs required. Software prices are high because there is little competition in this market. And there is a limited number of buyers who need to communicate between any two specific CADD systems. Finally, most CADD vendors release at least one, and sometimes two, new versions of their software each year. Many releases include changes in graphic data format, so direct translation software can have a life of a less than a year.

Users may purchase software maintenance contracts. Like hardware maintenance, these generally charge a monthly fee to guarantee users that the software will be upgraded when either CADD system changes its data format. Users can purchase each updated version as it becomes available.

The second way to tackle graphic data translation is through neutral format translators. A neutral format is a nonproprietary graphic data format intended to facilitate transfer of graphic information between CADD systems. Documentation is made available to the public. One such format, the Initial Graphic Exchange Specification (IGES), is an ANSI (American National Standards Institute) standard, and documentation is available through the National Technical Information Service in Washington. Other neutral formats have been designed by specific CADD vendors to facilitate data exchange with their systems. The two most frequently used are Auto Desk's Drawing Interchange Format (DXF) and Intergraph's Standard Interchange Format (SIF). The neutral format most commonly used a few years ago was SIF, but DXF now appears to be more generally accepted, particularly among PC-based CADD users. (The other format in which graphic data are sometimes transferred between CADD systems is a plot format, typically CalComp or Hewlett-Packard.)

Neutral format translation requires two steps. First, the originator of the data—in this case, the surveyor—translates the graphic information from a CADD system's proprietary format into the neutral format. This is the format in which the data are delivered to the client. The client must then translate the data from the neutral format to a CADD system's proprietary format.

A major inconvenience of this approach is that it takes at least twice as long as direct translation. With a large survey, it can eat up time on both the surveyor's and the user's systems. Another problem is that users may need to purchase translation programs between the neutral format and their CADD systems if vendors do not provide them as part of the CADD software purchases.

Finally, programs do not always execute properly. This can be due to an error in the software or a mistake on the part of the user. Translation programs, whether direct or neutral format, are no exception. The added difficulty with the neutral format approach is that it is difficult to pinpoint where the failure occurred—at the surveyor's end or at the user's. The situation is particularly frustrating when a client who has had painful experiences with unsuccessful and costly graphic data translations may demand that the data be delivered in their CADD system's format.

New computer products are being made available every day. Often there is a trade-off between the enhanced degree of functionality in state-of-the-art software packages and their limitations in translation capability. If a software package proves to be truly exceptional and finds a large number of users, translation software will almost surely follow. If the software has limited appeal, either because it is extremely special-purpose or because it is not well-marketed, compatibility problems will most likely persist.

Requests for digital data deliverables will certainly become more frequent. Large users like the U.S. Army Corps of Engineers have recently made major commitments to move to a computer-based design and documentation process. This means that not only will the Corps be requesting CADD deliverables, but increasing numbers of consultants will convert their operations to CADD to be able to satisfy the Corps' requirements. Surveyors currently looking at new computer systems or considering an upgrade should make data exchange capability a major criterion. They should contact major clients to determine their CADD preference and quiz prospective software vendors about their translation software capabilities. If the necessary translation software is available but too expensive, the vendor may be able to recommend service bureaus that provide translation services.

Surveyors who have computer systems and are generally pleased with their software's functionality, but are encountering requests for digital deliverables, should do a quick survey of their major clients to determine what CADD equipment they are using. They should then contact their software vendor to see what solutions they suggest. There may, in fact, be a translation program already available, through either vendor or a third party. If not, the more requests the vendor gets for translation capability, the more viable a translation program will appear as a new software product.

Another good source of information is a software users group, if one exists. Finally, the surveyor can contact CADD service bureaus in the area to see which data translation services they are able to perform. Fortunately, many service bureaus have invested heavily in translation software and are becoming expert in CADD data translation into a number of formats.

One caution: Be sure to test data translation software using "real life" data. Also ask for references who are surveyors or civil engineers. Graphic data translation is tricky, and a translation program that works wonderfully for 2D architectural floor plans may be totally incapable of handling 3D survey data.

A final concern in the delivery of digital data is the media on which the data will be transferred. CADD files are relatively large and extremely cumbersome to transfer via modem. Much preferable and more reliable is the physical transfer of a diskette, magnetic tape, or tape cartridge. When surveyors discuss CADD deliverables with clients, they should explore the question of which media the clients use. Those with large computers probably prefer 0.5-in. 9-track magnetic tape. Those using microcomputers will want diskettes or cartridge tapes. Whereas the large magnetic tape specification is standard, both diskettes and cartridges come in a variety of sizes and formats: high density, double density, etc. The surveyor may decide to forgo a 3.5-in. floppy drive if existing clients use 5.25-in. high-density diskettes.

Figure 5-7 depicts the basic requirements of a generic data collector.

## 5-8. Data Collector Requirements

The data collector is vital to large surveys using the total station. Assumptions or oversights made at the time of equipment purchase can force a survey operation into equipment problems on the job for the economic life of the equipment. Below are listed some options to consider for the data collector:

- Weatherproof, rugged/durable, designed for field use
- Nonvolatile memory ensures data safety.
- Allow the storage of at least 1,000 points
- Full search and edit routines immediately on the spot
- Automatic recording with electronic theodolites
- Manual entry and recording capability with the hardware that measures angles and distances
- Formatting must be very flexible for manual entry, even for various CADD leveling tasks.
- Capability to use two files in the collector: one file for collection, the other file for processed data for stakeout tasks

- Data collector must communicate with the electronic theodolite.
- All the features of the total station should be usable with the data collector purchased.
- The data collector must be compatible with the software you purchase or plan to use.
- Mixing brands should not cause a service problem.

## 5-9. Coding Field Data

Whether data are recorded manually or electronically, one of the most time-consuming survey operations is the recording of a code or description to properly identify the point during processing. For example, in a topographic or planimetric survey, identification points that locate the positions of curbs, gutters, center lines, manholes, and other similar features are essential for their correct plotting and contour interpolation.

Especially in topographic or planimetric surveying, many surveyors have wished for some way to speed up the process. For the most part, surveyors tolerate the time-consuming coding process because it is the only way of ensuring an accurate final product.

Despite this slow coding process when using data collectors available today, the advantages heavily outweigh the disadvantages. These advantages include collection blunder-free numeric data from electronic total stations virtually at the instant they are available and the error-free transfer of these data to an office computer system without the need for manual entry.

Field coding allows the crew to become the drafter and provide a more logical approach, as the field crew can virtually produce the map from the field data and eliminate the need for many field book sketches. They can also eliminate office plotting, editing by connecting the dots, etc., to produce a final product. The coding scheme is designed so the computer can interpret the recorded data without ambiguity to create a virtually finished product.

Although most of the codes required for survey operations will be found in the following pages, from time to time, additional codes may be required.

Either numerical point codes or alphanumeric point codes can be entered into the total station. This identification will vary from district to district, but the descriptor should be standardized throughout the Corps of Engineers.

Whenever districts require specialized point codes, then the attribute file may be edited to include these changes.

DATA COLLECTION PROCEDURES FOR THE TOTAL STATION

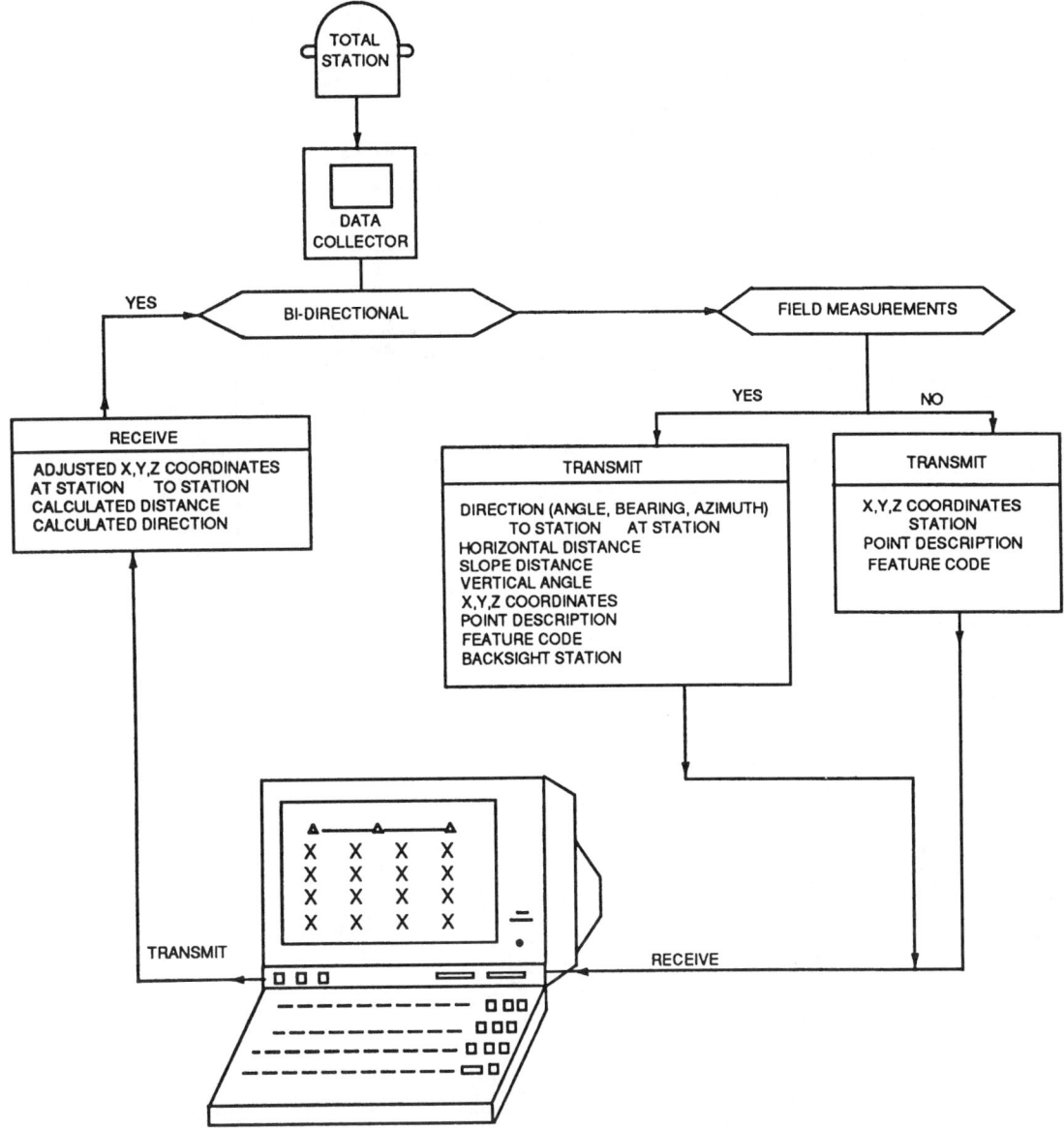

Figure 5-7. Functions of Generic Data Collector

## 5-10. Summary of Total Station FIELD-TO-FINISH Procedures

1. Gather field data and code the information.
2. Offload the data to computer, and process the information using equipment-specific software.
3. Create the ASCII Coordinate File containing point number, X-coordinate, Y-coordinate, Z-coordinate, standardized descriptor, and any additional notes.
4. Import the ASCII coordinate file into a CADD program and create a graphics file.
5. Use the CADD program to develop a final map with topographic, planimetric information that includes contours, utility information, etc.
6. Edit map.
7. Plot map.

This procedure is illustrated in Figure 5-8.

## 5-11. Data Collectors

### a. Geodimeter 126/400 Series

Geodimeter has integrated most, if not all, of the features of its established Geodot 126 series data collector into the 400 series. The 126 data collector is

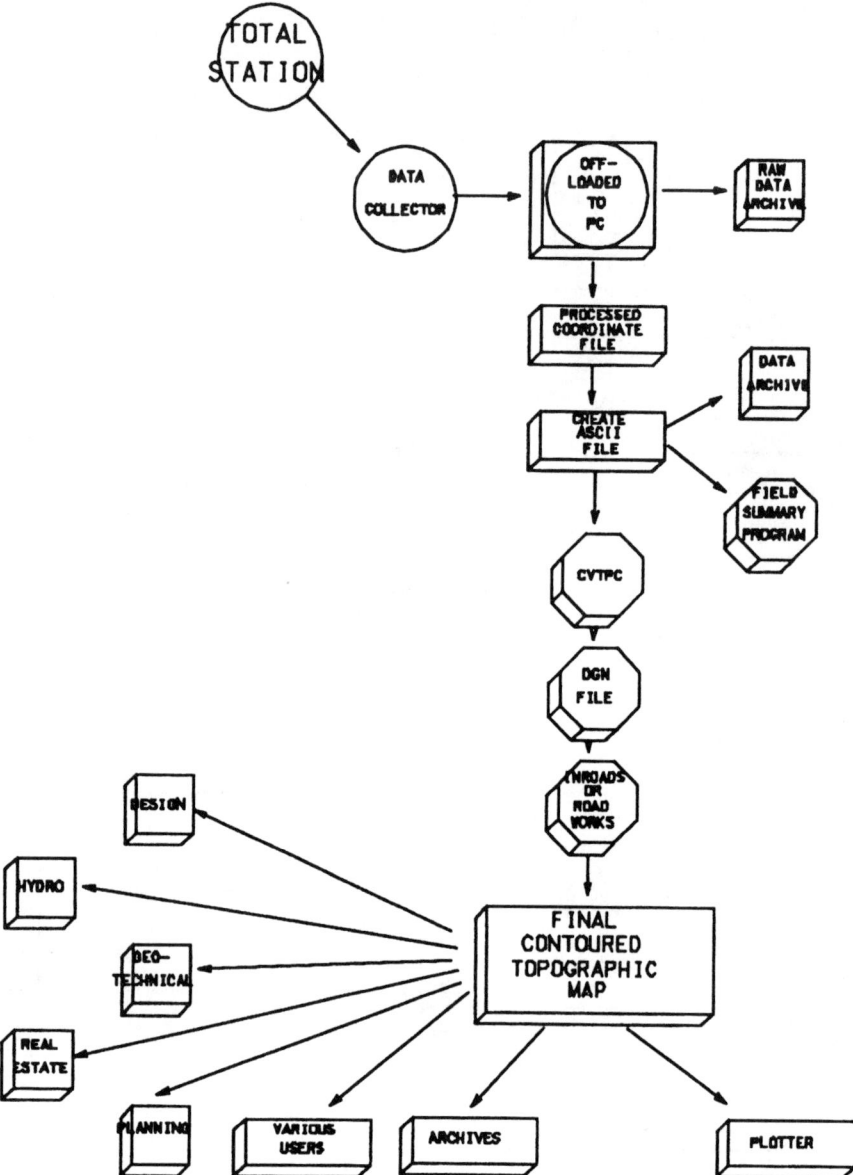

**Figure 5-8. Data Flow Process for Mapping**

still available for those users that may have older Geodimeter equipment.

- The power source for the Geodot 126 is four nickel–cadmium rechargeable batteries, which give a life of 15–20 hours between charges. The 400 series uses the on-board power supply of the instrument. Data integrity is provided by a lithium battery.
- The systems will operate at temperatures between 15 and 122 °F, which is somewhat higher than most systems and should be considered by those working in the northern areas.
- Data storage is described as 900 points for the 400 and 1,500 points for the 126. Both systems allow for additional (optional) memory. Both systems also allow data to be downloaded to an external data storage device through the HPIL that is integral to both systems.
- The Geodot 400 and the Geodot 126 will only support Geodimeter systems.
- Both systems support a number of calculating functions as well as northings, eastings, and elevations of points measured in the field.

### b. Lietz SDR Series

The Lietz SDR series of data collectors has a number of features that may be of interest.

- The system uses four AA batteries. The operating life is 120 hours. Data integrity is protected should the battery run down.
- The SDRs can operate in a temperature range of 4–122 °F.
- The data storage capacity is 32K for the SDR 20, 64K for the SDR 22, and 128K for the SDR 24.
- The SDR 24 will support total stations other than Lietz.
- Built-in programs handle many coordinate geometry functions.
- The data collector is programmable. These programs must be downloaded from the computer.

### c. Topcon FC-4

- The Topcon FC-4 contains an on-board rechargeable nickel–cadmium battery with sufficient power to operate the unit for 8 hours. The data are protected with a lithium battery that will assure the integrity of data for up to 6 months. The FC-4 will allow the use of an external battery when operating for extended periods.
- The system will operate at temperatures from −4 to 122 °F. With the optional water and dust cover, the unit is waterproof.
- Data storage capacity is 256K with additional storage available for disk and tape. In addition, a unique RAM disk makes it convenient for the field crews to store excess data.
- The FC-4 will support 39 models of instruments in the Topcon line and has the capability to support instruments of other manufacturers.
- The FC-4 has the common complement of coordinate geometry functions.
- The data collector is programmable. User-defined prompts can be input from the data collector keyboard. Custom programs can be done on the computer and downloaded to the FC-4. The FC-4 does not allow for programming from the keyboard.
- Topcon no longer manufactures the FC-1 and the PROPAC HA3 data collectors, but these data collectors remain in the inventory of many satisfied users.

### d. Wild GRE and REC Series

- The GRE system uses rechargeable nickel batteries and can be powered from external batteries. When used with an external power source, the GRE can run with the on-board battery removed. The REC module operates from the instrument battery. Data integrity is preserved with a lithium battery contained in the REC module.
- The systems can operate at temperatures that range from 4 to 122 °F. They are "splashproof."
- The data storage capacity of the current line of GRE data collectors is 128K. The older GRE 3 models range from 16K to 128K. The REC module is a 16K module.
- The GRE and GRM 10 will support only the Wild instruments.
- Built-in programs handle many coordinate geometry functions.
- The data collector is programmable. For all practical purposes, these programs must be completed on a PC and downloaded to the data collector.

# CHAPTER 6

# SURVEYOR DATA COLLECTOR INTERFACE AND FORMATS

For many surveying operations, electronic data collection is routine. However, after the data are collected, most software systems require a large amount of post-processing to produce a map showing planimetrics and contours.

## 6-1. Computer Interfacing

Many of the benefits of automated data collection are lost if the data stored cannot be automatically transferred to a computer system. The information is downloaded from the data collector to the computer. The file is usually an American Standard Code for Information Interface (ASCII) file. The measured distances, angles, or elevations entered into the data collector are output, but the format is not easily read. This is because the data are contiguous; nothing separates one piece of data from another. Every data collector stores data in a different format and the problem is to translate the data from the data collector format into a file with a standard ASCII format. Data standardization will become more important in the future, and surveyors should be searching for methods that make system integration easier.

## 6-2. Data Standardization

Data standardization is becoming more important as the concerns of redundant data and lack of shared data increase. Such standardization will reduce the cost of preparing the raw survey data for submission.

## 6-3. Coordinate File Coding

This section describes a coding scheme that can adequately define the survey parameters. These code records would be inserted into the ASCII coordinate file produced by the data collector. The codes are developed for general U.S. Army Corps of Engineers (USACE) ground topographic (topo) survey requirements for Architect-Engineer (A-E) applications. Additional codes may need to be developed to suit particular applications. All code records will begin with a "#" in column 1 and are limited to 80 columns. All comment records will begin with a ";" in column 1 and also are limited to 80 columns.

### a. Survey Job Parameters

Header records are required to describe the survey job parameters such as Horizontal Datum, Units of Measure, Survey Date, Job Location, Survey Firm, etc. H20 to H29 are reserved for job title. H30 to H99 are reserved for any comments about the survey job. If sketches or survey data were recorded in survey field books, the book and page numbers shall be indicated on the H12 and H13 records. This allows for an easy reference to original field data.

(H-RECORDS)
#H01  ASCII FILE NAME
#H02  SURVEY DATE
#H03  SURVEY ORDER
#H04  HORZ DATUM
#H05  JOB NUMBER
#H06  UNITS OF MEASURE
#H07  MAP PROJECTION
#H08  LOCATION
#H09  AE CONTRACTOR
#H10  BOOK NUMBER
#H11  PAGE NUMBER
#H12  COMBINED SCALE FACTOR
#H20  JOB TITLE
#H21  TITLE CONTINUATION
#H29  TITLE CONTINUATION
#H30  COMMENTS

#H31 COMMENTS CONTINUATION
#H99 COMMENTS CONTINUATION

### b. Horizontal/Vertical Control

All control points, whether found or established, must be described by control code records. Vertical control records are required to define the parameters such as Vertical Datum, Benchmark Name, Epoch, etc. used to determine the survey point elevations. These records are required at the beginning of a file and where the vertical parameters change.

(V-RECORDS)
#V01 VERTICAL BENCHMARK
#V02 GIVEN BM ELEVATION
#V03 EPOCH (YR OF ADJUSTMENT)
#V04 VERTICAL DATUM
#V05 CONDITION OF MARK
#V06 ELEVATION FOUND
#V10 ANY COMMENTS THAT THE
#V11 SURVEYOR MAY WANT TO MAKE
#V12 ABOUT THE BENCHMARK OR
#V99 LEVEL RUN

### c. Baseline Parameters

These records describe the reference baseline. If a baseline listing is available on diskette the user may include the file name on the B00 record. Each baseline point of intersection (PI) is defined by its coordinates and station number. Curve data are defined by BC, BI, and BT records. These records define the coordinates and station number of the point of curvature, point of intersection, and point of tangency (PC, PI, and PT, respectively).

(B-RECORDS)
#B01 X-crd Y-crd STATION
#B02 X-crd Y-crd STATION
#B99 X-crd Y-crd STATION

BASELINE FILE NAME
#B00 FILE.EXT

CURVE PARAMETERS
#BC1 X-crd Y-crd STA (PC)
#BI1 X-crd Y-crd STA (PI)
#BT1 X-crd Y-crd STA (PT)

### d. Temporary Benchmarks (TBMs)

All TBMs used, whether established or found, must be defined with TBM records. The PBM used to set the TBM will be assumed to be the previous V01 record (V-Records). The date set will come from the last "H02" record. T10 through T99 are used for description of mark.

(T-RECORDS)
#T01 TBM NAME
#T02 GIVEN ELEVATION
#T05 CONDITION OF MARK
#T06 ELEVATION FOUND
#T10 DESCRIPTION
#T11 DESCRIPTION CONTINUED
#T99 DESCRIPTION CONTINUED

### e. Water Surface Elevation

Gauge records are required each time a gauge is read.

(G-RECORDS)
#G01 STAFF GAUGE CODE#
#G02 GAUGE NAME
#G03 GAUGE READING
#G04 TIME OF READING
#G10 DESCRIPTIONS AND OR
#G11 COMMENTS ARE INCLUDED
#G12 FROM G10 TO G99

### f. Cross-Section Parameters

Each cross section must be preceded by an X01 record. If the section contains sounding data controlled by a gauge, X03 (time) and X04 (elevation) records must be included showing the interpolated water surface elevation.

(X-RECORDS)
#X01 BX BY EX EY STATION NAME
#X02 RANGE CODE
#X03 TIME OF SECTION (IF SNG)
#X04 WATER SURFACE ELEVATION

### g. Profile Parameters

Each reach of profile must be preceded by a P01 record. If the profile contains sounding data controlled by a gauge, a P03 (time) and P04 (elevation) record must be included showing the interpolated water surface elevation.

(P-RECORDS)
#P01 BX BY STATION
#P03 TIME OF PROFILE (IF SNG)
#P04 WATER SURFACE ELEVATION
#P10 TITLE OF PROFILE

### h. Miscellaneous Records

These records are required on miscellaneous shots. The record will contain a general description of the points that follow.

(M-RECORDS)

#M01  
#M02 Borehole locations at the south end of the ammo plant located in the U.S. Army Reserve Complex in Corn Bayou, La.,  
#M03 near the WABPL.

## 6-4. Data Sets

A data set is defined as a cross section, a profile, or a group of topo shots. A data set begins and ends with the M, P, or X code records. For example a profile data set begins with the P records and is terminated by any M, P, or X record.

INDEX OF CODE RECORDS

#B01 Coordinates and station of baseline PI  
#B00 Name of ASCII coordinate file that contains the survey data  
#BC1 Coordinates and station of point of curvature for curve #1  
#BT1 Coordinates and station of point of tangency for curve #1  
#BI2 Coordinates of point of intersection for curve #2  
#C01 Party Chief  
#C02 Instrument Person  
#C03 Rodperson  
#G01 Staff gauge code number supplied by USCOE  
#G02 Name of gauge  
#G03 Water surface elevation as read on gauge  
#G04 Time (1423) of gauge reading based on 24-hour clock  
#G10–G99 Descriptions and/or comments, limited to 75 characters per record  
#H01 Standard DOS file name of ASCII file which contains the survey data (more than one file is allowed per survey job)  
#H02 Date (DD/MM/YY) on which the following information was obtained  
#H03 Order (accuracy) of survey (1,2,3..AA)  
#H04 Horizontal datum on which the survey is referenced (NAD-1927, NAD-1983, WGS-84, ...)  
#H05 Job number of survey (YY–JJJ)  
#H06 Unit of linear measure (FT, MT, MI, ...)  
#H07 Map projection. Use standard list of projection codes (1702, 1703, ...).  
#H08 Location of survey such as nearest town, river, channel, basin (more than one location is allowed per survey)  
#H09 Survey firm or organization  
#H10 Index number of survey field book in which the following information is recorded  
#H11 Page number of field book specified by previous H10 code on which the following information is recorded  
#H12 Combined scale factor  
#H20 Title of survey job (limited to 75 characters per record)  
#H21–H29 Continuation of survey job title  
#H30 Reserved for any comments about the survey job (limited to 75 characters per record)  
#H31–H99 Continuation of comments about the survey job  
#I01 Instrument  
#I02 Serial number  
#M01–M99 Description of miscellaneous survey points that follow  
#P01 The profile segment's beginning $x,y$ coordinates and stationing  
#P03 Time of profile (needed only if elevations of points are relative to prorated water surface)  
#P04 Prorated water surface elevation used for elevation of points in profile  
#T01 Name of TBM  
#T02 Given elevation of TBM  
#T05 Condition of TBM  
#T06 Found elevation of TBM  
#T10–T99 Description of TBM  
#X01 The range line definition, which contains the end point coordinates, station, and name of the range  
#X02 Range code or index number  
#X03 Time of cross-section (needed only if elevations of points are relative to prorated water surface)  
#X04 Prorated water surface elevation used for elevation of points in cross section  
#W01 Temperature  
#W02 Pressure  
#W03 Humidity

#W04 Cloud conditions (0–10% = clear; 10–50% = scattered; 50–90% = broken; 90–100% = obscured)

#W05 Wind speed

#W06 Wind direction (N, S, E, W, NE, NW, SE, SW)

## 6-5. Computer-Aided Design and Drafting (CADD) Interface

CADD software packages are commonly available that can produce basic survey plots to finished map sheets. Such drafting tools offer the surveyor more accuracy, efficiency, flexibility, and quality in the production of hard copy plots. Microstation™, which was available through the Corps-wide CADD contract with Intergraph Corporation, is commonly available and used in USACE offices. However, numerous other CADD packages that run on PC-based system, such as AutoCAD™, GWN-COGO™, and TRU-CAD™, are available.

## 6-6. Total Station Data Collection and Input to CADD

Survey data can be entered into a CADD program by a variety of methods. The most favorable means is through digital data files produced by electronic survey equipment. Total stations, global positioning system (GPS) survey receivers, and some electronic levels are commonly capable of recording survey data on electronic data collectors, floppy disks, cassette tapes, magnetic cards, internal media, or interfaced field computers. Such logging of data greatly increases the efficiency and accuracy of data collection and eliminates human error associated with field book recording. These digital data files also eliminate the tedious and error-prone manual entry of data into CADD programs. It should be noted that automatic data logging clearly offers a superior method for recording and processing survey angles, ranges, or coordinates but does not eliminate the field book. Survey conditions, description of the project, unplanned procedures, and other pertinent information must always be recorded by field personnel to establish complete survey records.

For total station instruments, various software/hardware packages are available to collect and process survey data. For example, survey adjustment packages such as Wildsoft™, Pacsoft™, and SDRMAP™, which are PC-based, will interface to a variety of data collectors. Some collectors are actually PC-based processors that can log total station data and run various survey adjustment software packages. Intergraph's Electronic Theodolite Interface™ offers a full set of hardware and software to log survey data, perform post-processing and adjustments, and import the data into an Intergraph workstation for CADD processing.

If procuring components of a data collection and processing system, compatibility between components and a minimum capability must be assured. Survey coordinates with a descriptor or code to indicate the surveyed feature should be input, as a minimum, to the CADD system. ASCII X-Y-Z or latitude-longitude-height data, along with alphanumeric descriptor data, are usually accepted by CADD software and are commonly output by data collectors or survey processing programs. The CADD program should have some flexibility in the order the coordinates are received (i.e., X-Y-Z, Z-X-Y, etc.) and the length of the data records.

More complex and sophisticated information, such as contour lines and symbols, can sometimes be passed from survey to CADD programs through common graphic formats, such as Auto Desk's Drawing Interchange Format (DXF). However, note that a 100% reliable transfer of graphic data is not always possible. For example, contour lines passed to a CADD program in DXF format may have isolated breaks or overlap. Transfer of graphic data using proprietary formats is usually most reliable.

## 6-7. CADD Plotting

CADD systems offer extreme flexibility in data plotting and usually can follow the specification described in the previous section. Sheet sizes available depend on the plotter or printer. "A" size is usually available on all output devices, and in most cases, a desktop printer will suffice. Some desktop devices are capable of "B" and "C" sizes, and standing floor-mounted plotters are usually required for "D" and "E" sizes. Pen plotters can output on most desired media, including paper and Mylar.

Plotters that use mechanically guided pens are the most common, and usually inexpensive, plotting device. With the proper pen cartridges, the quality of plots is equal to or greater than that of professionally drafted manual plots. Note that these devices produce only line segments or curved lines. Thus, a shape that is color-filled would be produced through numerous color strokes. Such tedious plots require long plotting times and produce more wear on the pens and the plotter itself.

Plotters that use electrostatic, inkjet, thermal, or laser techniques are becoming more common. Such devices produce excellent-quality plots and are much quicker than pen plotters. However, these devices are considerably more expensive and may output only on specific media.

# CHAPTER 7

# MAP COMPILATION

## 7-1. General

This chapter describes a process, commonly used by most districts, to create Intergraph design (DGN) files from digital survey files. Note: Other computer-aided design and drafting (CADD) systems and procedures can be used, providing that the same coordinate information and descriptive data can be imported into the CADD routine.

## 7-2. CVTPC

The collected survey data are of little value in their present form. The coded ASCII Coordinate File that has been edited to conform to the U.S. Army Corps of Engineers (USACE) Coordinate File Standards is now ready to be converted into an Intergraph design file. This conversion is accomplished using the software package CVTPC, and the process shown in Figure 7-1. Figure 7-2 displays the screen in CVTPC. From this figure, all the data attributes can be seen. The attribute setup is constructed of rows and columns. For example, level, color, and weight are columns. Point, Line, and Elev. are rows. The rows are assigned to the columns selected. This becomes the file that is written. The only other requirement is the file from which the data are read. This file information is entered in the model setup section on the right side of the screen. The file information is entered into the Input Files block at the top of the screen. Documentation can be obtained from the U.S. Army Topographic Engineering Center.

CVTPC converts the ASCII Coordinate File into DGN files. The program code is written in Intergraph's Microstation Design Language (MDL) and therefore can only be executed while Microstation is running and the user is currently in a design file.

The ASCII Coordinate File may have its coordinates and point descriptors placed in any order, for example, "Point name *X Y* and *Z* point code" or "*Y X Z* point code, point name." The ASCII data can be separated by spaces or commas, or they can be designated by column position. For example, ASCII data files separated by spaces would be of the form

1 32987.34 45890.01 123.44 MANHOLE

Data separated by commas would be of the form

1,32987.34,45890.01,123.44, MANHOLE

Data separated by columns would be of the form

1 32987.34 45890.01 123.44 MANHOLE

where the point name (number) is in column 0, X-coordinate in columns 5–12, Y-coordinate in columns 15–22, Z-coordinate in columns 25–30, and the point code in columns 31–37. CVTPC will run on any PC that has Microstation version 4.0 or later. A

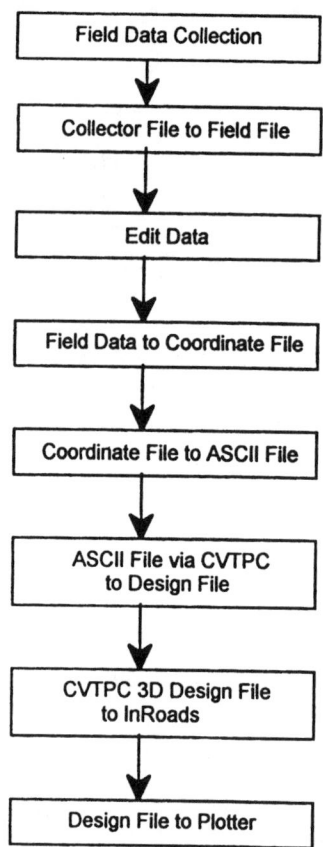

**Figure 7-1. Field-to-Map Conversion Process**

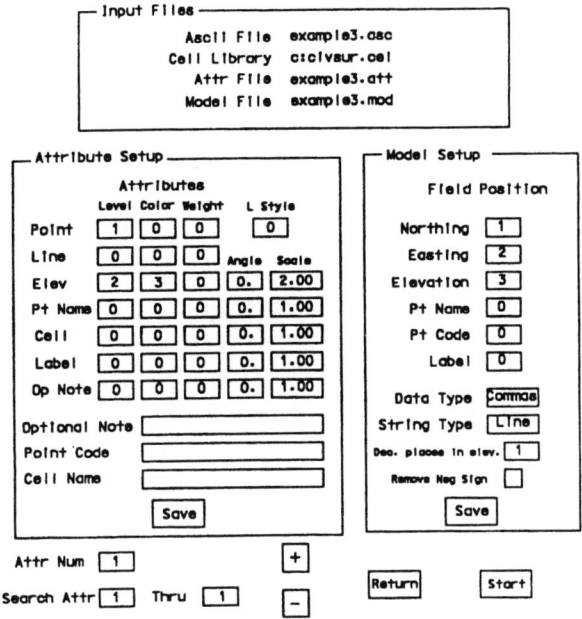

**Figure 7-2. CVTPC Screen**

typical ASCII file is shown in Figure 7-3. An example of the basic information for a typical utility is shown in Figure 7-4.

## 7-3. File Descriptions

- Attribute File: Links the point descriptors in *ASCII Coordinate File* to the cells, assigns colors, levels, weights, text scale, and active angle.
- Model File: Specifies seed file, cell library and contains a blueprint of the ASCII Coordinate File format. The method of coordinate input, column or comma.
- CIVSUR.cel: Cell Library containing Corps of Engineers standardized cells.
- ASCII Coordinate File: Input for CVTPC and the output of the coordinate file of the software specific to the particular instrument/data collector.

## 7-4. Overview of Topographic Survey Data Flow

Figure 7-5 outlines the various routes by which topographic data are processed to a final site plan map form. Note that this figure includes digital topographic data collected from different sensors (e.g., aerial, hydrographic sonar/acoustic). Figures 7-6, 7-7, and 7-8 depict USACE standard feature level assignments, USACE civil/site level and element symbology, and USACE surveying and mapping

```
1,782365.363465090,2275214.86945614,573.478884514979,U/G COMM BOX
2,782345.953423030,2275249.16042761,572.481197277167,CATCH BASIN
3,782308.161611151,2275267.51443841,573.374514576417,LIGHT POLE
4,782228.247274107,2275336.70541060,573.139822606539,LIGHT POLE
5,782227.355064638,2275263.13552007,573.451098999176,LIGHT POLE
6,782125.098941644,2275248.01129811,572.936901550331,LIGHT POLE
7,782102.301674787,2275250.72455093,572.629547106902,CHAIN LINK
8,782179.746636550,2275246.70661877,572.708123979877,FIRE HYDRANT
9,782224.871955784,2275248.70849014,573.515531194125,CHAIN LINK
10,782235.317587276,2275262.16604342,573.681561702655,PHONE BOOTH
11,782226.843634089,2275353.61702810,573.005122075631,CHAIN LINK
12,782278.987272298,2275210.02584227,573.622937299323,U/G CONDUIT
13,782278.462087678,2275218.36099477,573.385310919142,U/G CONDUIT
14,782377.724779433,2275216.13381257,573.689131881914,U/G CONDUIT
15,782317.005550843,2275231.21562848,573.110888677882,CONC CURB
16,782312.781072156,2275259.69891992,573.023770690619,CONC CURB
17,782304.025556457,2275277.96178023,572.995097334823,CONC CURB
18,782344.592211204,2275302.61925727,573.094892904533,CONC CURB
19,782351.947810373,2275284.63062089,573.100396667034,CONC CURB
20,782368.891819889,2275277.32543843,573.101642797183,CONC CURB
21,782413.436614815,2275277.02973580,573.602841655186,CONC CURB
22,782412.998966620,2275254.18584339,573.594048399250,CONC CURB
23,782364.369922795,2275254.93706052,573.110646664148,CONC CURB
24,782350.952703436,2275251.06562781,573.113947762143,CONC CURB
25,782339.514247482,2275230.68014576,573.110252423328,CONC CURB
26,782339.152169872,2275233.26204572,572.706989862464,TOE
27,782350.666626204,2275251.55522711,572.609987649321,TOE
28,782365.601261555,2275255.43408859,572.649109806971,TOE
29,782390.646837720,2275254.88205653,572.886679217580,TOE
30,782394.667870564,2275276.66020279,572.930115869840,TOE
31,782370.428136512,2275277.12200200,572.765204178545,TOE
32,782346.715042993,2275291.00881484,572.702835922715,TOE
33,782344.103660145,2275305.92367362,572.561707170088,TOE
34,782285.410243099,2275310.02658394,572.336831930867,TOE
35,782302.939035655,2275280.70870612,572.599648988195,TOE
36,782314.363944258,2275255.65543457,572.633969810851,TOE
37,782317.284924185,2275233.33727174,572.772807520655,TOE
38,782293.653667242,2275168.80899235,573.957935950842,BLDG CORNER
39,782277.733086761,2275169.25226746,573.838209222809,BLDG CORNER
40,782278.411855050,2275218.47606015,573.285363627969,BLDG CORNER
41,782265.820393806,2275218.95538862,576.099082375093,BLDG CORNER
42,782265.827976941,2275224.22263680,573.095115769811,BLDG CORNER
43,782254.339003791,2275224.48758308,573.133398941741,BLDG CORNER
44,782241.692493278,2275219.20328765,573.160643415879,BLDG CORNER
45,782279.469604898,2275301.11452594,573.048773751455,DECIDUOUS
46,782294.991196529,2275275.08351122,573.769161734049,DECIDUOUS
47,782354.525110295,2275119.50828985,575.426280489799,DECIDUOUS
48,782354.422586876,2275144.77453043,574.911439829978,DECIDUOUS
49,782355.489879412,2275183.19940609,574.455544604298,DECIDUOUS
50,782355.961459436,2275216.40162001,574.109218498818,DECIDUOUS
51,782388.754534972,2275241.20230956,574.274163732338,DECIDUOUS
52,782389.492289031,2275289.97462803,573.805673749121,DECIDUOUS
53,782366.256525734,2275244.95355959,574.408170510696,DECIDUOUS
54,782339.168426240,2275209.68225495,572.833010738988,EDGE RD
55,782339.149044318,2275162.37332474,573.046984674136,EDGE RD
56,782348.647664815,2275112.85519829,573.897230805704,ASPHALT
57,782349.792486408,2275159.95969757,573.704089469543,ASPHALT
```

**Figure 7-3. Typical ASCII File**

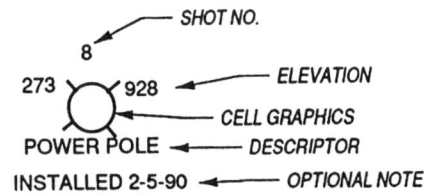

**Figure 7-4. Information Provided by Field Shot**

level and element symbology, respectively. Consult the USACE CADD Manual for details.

## 7-5. Typical Point Descriptors Used in Topographic Surveying

### a. Control

2X2 HUB/TACK
PK NAIL
RR SPIKE
PIPE
REBAR
1X2 STAKE
BRASS CAP

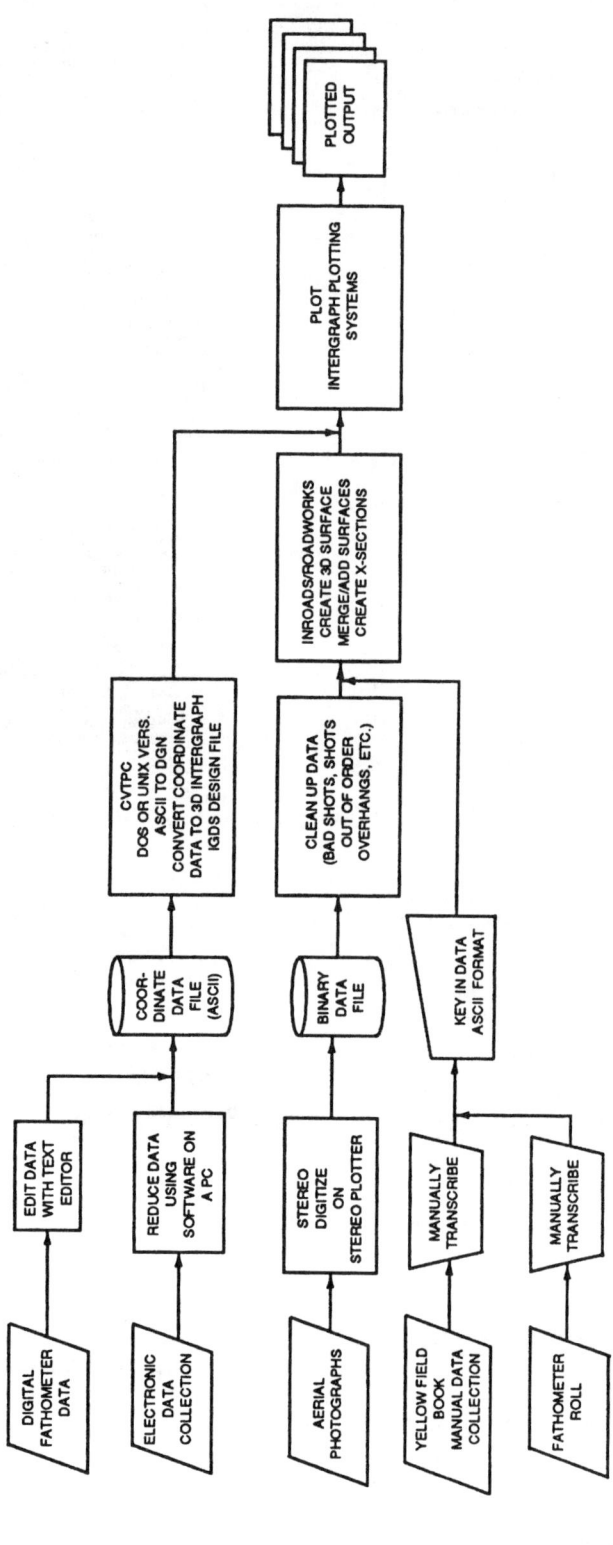

**Figure 7-5. Overview of Survey Data Flow**

# MAP COMPILATION

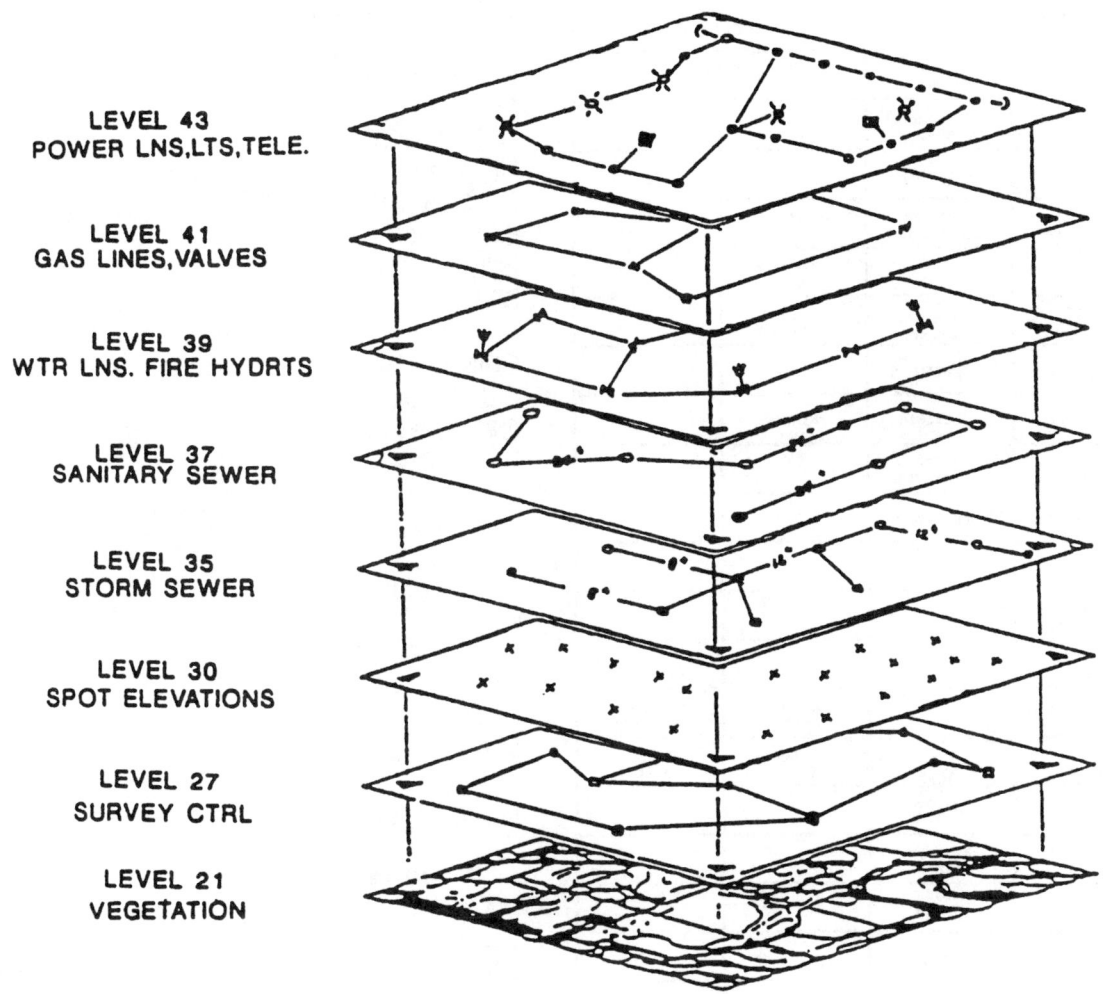

**Figure 7-6. Feature Level Assignments**

NAIL
FILED X
HAND DRILL HOLE
CHISLED X
BOLT
REBAR/CAP
MON
ALUM MON
CONC MON
COE MON
NGS MON
USGS MON
GPS MON
REFERENCE POINT
PICTURE POINT
SIXTEENTH CORNER
SECTION CORNER
QUARTER CORNER
CLOSING CORNER
MEANDER CORNER
WITNESS CORNER

HOMESTEAD CORNER
BENCHMARK
TBM
PBM

**b. Topographic**
GROUND SHOT
BLDG CORNER
U/G CABLE MARKER
CURB
PAD
ASPHALT
TOP/RIPRAP
TOE/RIPRAP
SLOPE
WE/WS
TOP
CROWN
TOE
C/L
B/L

| PLANIMETRIC | | MISCELLANEOUS | |
|---|---|---|---|
| GEOMETRY | TEXT | | |
| | | 1 SHEET DEPENDENT INFO | |
| 2 COORDINATE GRID | 3 COORDINATE GRID | | |
| 4 BUILDINGS | 5 BUILDINGS | | |
| 6 ROADS, RR CENTERLINES | 7 ROADS, RR CENTERLINES | 8 ROADS, RR SIDEWALKS | 9 CONCRETE JOINTS |
| | | 10 CONCRETE JOINT ELEV | |
| 11 RUNWAY TAXIWAY APRONS | 12 RUNWAY TAXIWAY APRONS | 13 PAVEMENT MARKINGS | |
| 14 STRUCTURES | 15 STRUCTURES | | |
| 16 CULVERTS | 17 CULVERTS | 18 RIPRAP | |
| 19 WATER FEATURES | 20 WATER FEATURES | | |
| 21 VEGETATION | 22 VEGETATION | | |
| 23 FENCE | 24 FENCE | | |
| 25 BOUNDARIES CADASTRAL | 26 BOUNDARIES CADASTRAL | | |
| 27 CONTROL POINTS | 28 CONTROL POINTS | 29 BREAKLINE | 30 SPOT ELEVATIONS |
| | | 31 MAJOR CONTOUR | 32 CONTOUR ANNOTATION |
| UTILITIES | | 33 MINOR CONTOUR | 34 BORE HOLES & TEXT |
| GEOMETRY | TEXT | | |
| 35 STORM SEWER | 36 STORM SEWER | | |
| 37 SANITARY SEWER | 38 SANITARY SEWER | | |
| 39 WATER | 40 WATER | | |
| 41 GAS | 42 GAS | | |
| 43 ELECTRICAL TELEPHONE | 44 ELECTRICAL TELEPHONE | | |
| 45 STEAM | 46 STEAM | 47 CROSS-SECTION T PROFILES | 48 DETAILS INSERTS |
| | | 49 SOUNDINGS | |
| 50 CHANNEL LINES DISP. AREAS | 51 CHANNEL LINES DISP. AREAS | 52 NAVIGAT. AID & TEXT | 53 LEVEES, DIKES & TEXT |
| 54 PIPELINES STRUCTURES | 55 PIPELINES STRUCTURES | 56 STATIONING & MILE MARKERS | 57 REVETMENTS & TEXT |
| | | 58 VESSEL TRACK LINE | 59 BORDER N. ARROW LEGEND |
| | | 60 | 61 |
| | | 62 | 63 PROJECT SUMMARY REPORT |

CIVSUR.CEL

**Figure 7-7. USACE Civil/Site Levels and Element Symbology**

DRIVEWAY
SIDEWALK
PORCH
ROCK
STEPS
AC@PCC JOINT
CUT-OFF FENCE POST
STEEL GUARD POST
WOOD GUARD POST
PAINT STRIPE
RETAINING WALL
SIGN
TOP STRUCTURE
HEADWALL
FLOODWALL
EXPANSION JOINT
BUILDING LINE
DRAIN
TREE LINE
CONIFEROUS
DECIDUOUS
ORNAMENTAL
HAZARDOUS WASTE VAULT
C/L RR TRACK
TOP OF RAIL
O/H PIPE
CHAIN LINK
BARBED WIRE
WOOD FENCE
GROUNDING ROD
AIRCRAFT TIE-DOWN
ASPHALT PATCH
CONC. PATCH
TOWER LEG

**c. Road Work**
SHOULDER ROAD
C/L ROAD
EDGE ROAD
RAMP

**d. Bridges**
END OF BRIDGE
PIER TOP
PIER TOE
ARCH START
ARCH CREST
ARCH END
ARCH TOE

**e. Electrical**
U/G STREET LIGHTING BOX
TRAFFIC SIG CONTROL BOX
ELECTRICAL OUTLET
BREAKER BOK
SWITCH BOX
ELECTRICAL VAULT
ELECTRICAL SPLICE
O/H POWER LINE
POWER POLE
POWER POLE/TRANS
GUY POLE
DOWN GUY
U/G CONDUIT

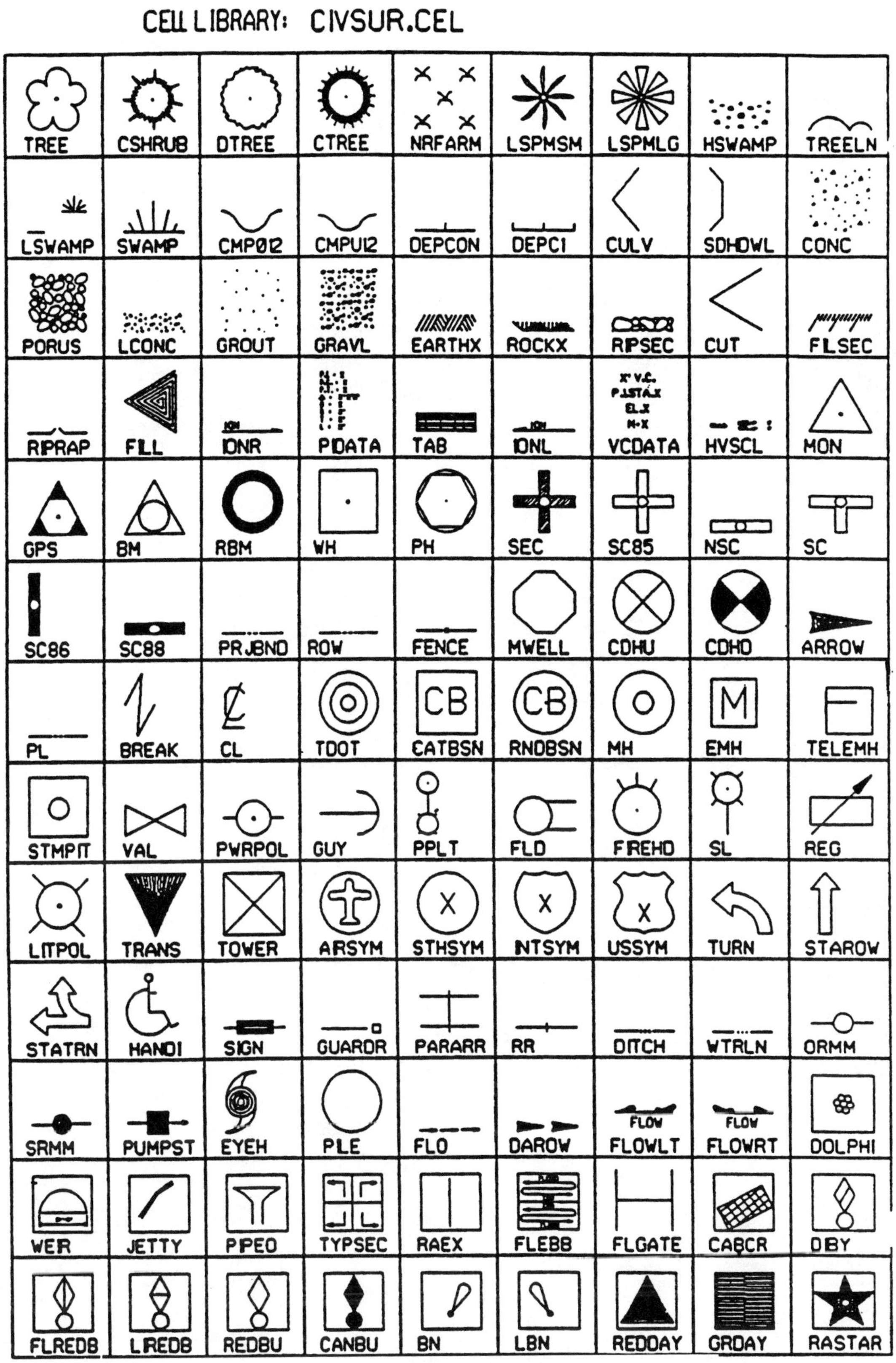

MAP COMPILATION

CELL LIBRARY: CIVSUR.CEL

| | | | | | | | | |
|---|---|---|---|---|---|---|---|---|
| TREE | CSHRUB | DTREE | CTREE | NRFARM | LSPMSM | LSPMLG | HSWAMP | TREELN |
| LSWAMP | SWAMP | CMP012 | CMPU12 | DEPCON | DEPC1 | CULV | SDHOWL | CONC |
| PORUS | LCONC | GROUT | GRAVL | EARTHX | ROCKX | RIPSEC | CUT | FLSEC |
| RIPRAP | FILL | IDNR | PIDATA | TAB | IDNL | VCDATA | HVSCL | MON |
| GPS | BM | RBM | WH | PH | SEC | SC85 | NSC | SC |
| SC86 | SC88 | PRJBND | ROW | FENCE | MWELL | CDHU | CDHD | ARROW |
| PL | BREAK | CL | TDOT | CATBSN | RNDBSN | MH | EMH | TELEMH |
| STMPIT | VAL | PWRPOL | GUY | PPLT | FLD | FREHD | SL | REG |
| LITPOL | TRANS | TOWER | ARSYM | STHSYM | NTSYM | USSYM | TURN | STAROW |
| STATRN | HANDI | SIGN | GUARDR | PARARR | RR | DITCH | WTRLN | ORMM |
| SRMM | PUMPST | EYEH | PLE | FLO | DAROW | FLOWLT | FLOWRT | DOLPHI |
| WER | JETTY | PIPEO | TYPSEC | RAEX | FLEBB | FLGATE | CABCR | DBY |
| FLREDB | LREDB | REDBU | CANBU | BN | LBN | REDDAY | GRDAY | RASTAR |

**Figure 7-8. USACE Surveying and Mapping Level and Element Symbology**

LIGHT POLE
ELECTRICAL MH
TRANSFORMER
RWY LIGHT
POWER @ BLDG

### f. Sanitary
SANITARY MH
SANITARY CLEANOUT
SANITARY LINE
EXPOSED SEWER PIPE

### g. Storm
TOP CONC DRAIN TROUGH
TOE CONC DRAIN TROUGH
GRATED STORM MH
STORM MH
CATCH BASIN
DRAIN PIT
CULVERT
FLOW DITCH
CURTAIN DRAIN

### h. Water
A/G WATER VALVE
AIR RELIEF VALVE
RISER
WATER MH
WATER LINE
SPRINKLER CONTROL VALVE
FIRE HYDRANT
WATER VALVE
WATER METER
WATER STANDPIPE

### i. Communication
TELEPHONE POLE
PHONE BOOTH
FIRE ALARM
U/G COMM BOX
TELEPHONE MH
U/G COMM

O/H COMM
TELE SPLICE BOX
TELE @ BLDG
CABLE TV

### j. Fuel
FILLER PIPE U/G TANK
A/G FUEL TANK
FUEL TANK VENT PIPE
FUEL PIT

### k. Gas
GAS PAINT MARK
GAS VALVE
GAS METER
GAS LINE

### l. Heating
STEAM PIT
STEAM MH
A/G STEAM LINE

### m. Geotech
PVC PIEZOMETER
PVC SLOPE INDICATOR
TEST WELL
MONITORING DEVICE
CDHU
CDHD
WELL HEAD

### n. Hydro
BUOY
DOLPHIN
PILING
SOUNDING

### o. Breaklines
C/L ROAD
EDGE ROAD
SLOPE BREAK
BREAKLINE

# CHAPTER 8

# ARCHITECT-ENGINEER CONTRACTS

## 8-1. General

This chapter presents an overview for preparing site plan mapping scopes of work for professional surveying and mapping services performed by Architect-Engineer (A-E) contract labor forces. Preparation of a statement of work and cost estimates is necessary to properly negotiate A-E contracted services. These services are either obtained through an indefinite delivery order surveying services contract or scheduled within A-E design services. The cost estimate should reflect all significant cost phases related to the mapping project.

## 8-2. Preparation

The preparation of a scope of work and cost estimate will depend largely on the intended use of the final mapping product. Table 2-1 defines target scales, contour intervals, and accuracy requirements associated with the various engineering mapping projects performed in the U.S. Army Corps of Engineers (USACE). Site conditions, project location, existing area control, mapping limits, available supporting data, instrumentation and procedures, transportation, personnel, equipment, deliverables, procedures, and individual disciplines required to complete the job are just some of the contributing factors that need to be properly defined.

## 8-3. Scope of Work

The scope of work should define the specific project requirements and deliverable items of work to be accomplished under the delivery order. Supporting guide specifications should define general contract and technical specifications that normally would pertain to intended use. The scope of work should contain the following sections:

- general statement and location of work
- specific requirements related to the site and intended use of the mapping product
- deliverable items of work
- field procedures and standards
- special instructions
- data and materials to be furnished for the contractors use
- completion schedules

A sample scope of work for a detailed site plan mapping project is shown in Appendix E. This scope of work would typically be used on a delivery order against a basic mapping contract.

# CHAPTER 9

# ROUTE SURVEYING

## 9-1. General

Route surveys are most commonly used for levees, stream channels, highways, railways, canals, power transmission lines, pipelines, and other utilities. In general, route surveys consist of

- determining ground configuration and the location of objects within and along a proposed route.
- establishing the alignment of the route.
- determining volumes of earthwork required for construction.

After the initial staking of the alignment has been closed through a set of primary control points and adjustments have been made, centerline/baseline stationing will identify all points established on the route. Differential levels are established through the area from two benchmarks previously established. Cross sections in the past were taken left and right of centerline. Today, digital terrain models (DTMs) or photogrammetry is used to produce cross sections for design grades. Surveys may be conducted to check these sections at intermittent stations along the centerline. Ground elevations and features will be recorded as required.

## 9-2. Horizontal Circular Curves

Route surveys often require layout of horizontal curves. The point of curvature (PC), point of intersection (PI), and point of tangency (PT) should be established on centerline, identified, and staked, including offsets to the centerline. Figure 9-1 is a sketch of a horizontal curve. The traverse routes through the curve will be included into the closed traverse through two primary or secondary control points for closure and adjustment. Field layout/stakeout should be no more than 100 ft along the curve on even stationing. PCs, angle points, and/or PIs should be referenced (line-of-sight) outside the clearing limits or the construction area.

## 9-3. Deflection Angles

The angles formed between the back tangent and a line from the PC to a point on the curve is the deflection angle to the curve. The deflection to the point on the curve is given by the equation:

$\Delta$ = arc length/radius

where

$\Delta$ = deflection angle or central angle

arc length = arc length found by subtracting the station number

radius = distance from the radius point to the centerline of the right-of-way alignment

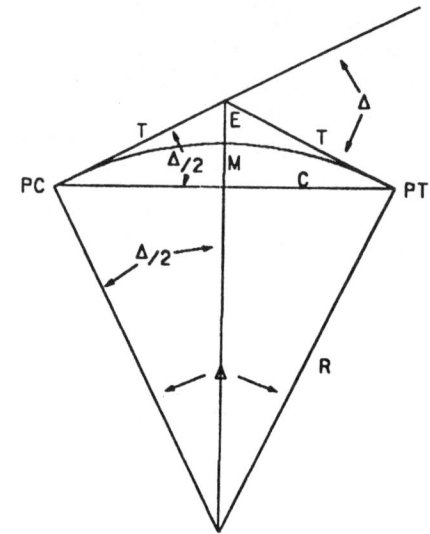

where

$\Delta$ = Central and Deflection Angle
$PC$ = Point of Curvature
$PT$ = Point of Tangency
$T$ = Tangent
$R$ = Radius
$C$ = Long Chord
$M$ = Middle Ordinate
$E$ = External

**Figure 9-1. Horizontal Curve**

# ROUTE SURVEYING

Transportation engineers compute minimum/maximum allowable curvature based on weight and speed. Surveyors fit curves to existing alignments based on the minimum/maximum curvature limits. Curvature limits apply only to primary roads. Secondary roads or roads where speeds are low are based on radius and deflection angle. Two kinds of formulas have been used to fit curves based solely on curvature:

- degree of curvature—arc definition
- degree of curvature—chord definition

## 9-4. Degree of Curve—Arc Definition

The standard 100-ft steel highway chain used by surveyors was the basis of the amount of curvature developed in 100 ft of arc. The ratio of curvature over a 100-ft arc is equal to the total degrees in a circle over the total arc length in a circle.

$$D_C/100 = 360°/(2 \times \pi \times R)$$

or

$$D_C = 100' \text{ arc} \times 1/R \times 180/\pi$$

Usually, the curvature will be specified. The surveyor needs to solve for the radius. Rearranging the above formula yields

$$R = 5729.578/D_C$$

The computed radius and directions (azimuths or bearings) of the straight portions of the right-of-way, called tangents, are usually used to compute the curve in the field.

## 9-5. Degree of Curve—Chord Definition

The chord definition was popular in the railroad industry. Some U.S. Army Corps of Engineers (USACE) districts use this method. The definition is valid because the curvature is slight in railroad curves and the difference between 100 ft of arc and 100 ft of chord cannot be measured with a steel chain to the nearest 0.01 ft. The formula for any chord is

$$\text{Chord} = 2 \times R \times \sin(\Delta/2)$$

The method defines the amount of curvature found in a 100-ft chord. Substituting the value of 100 into the standard chord formula and rearranging gives the degree of curve—chord definition:

$$\sin(D_C/2) = 50/R$$

where $50 = 100/2$.

## 9-6. Curve Stakeouts

The first and most important point in a curve stakeout is the PI of two tangent sections of a right-of-way. This point is set in lieu of the radius because the radius may be too far away from an instrument station for curves with small curvature. All PIs are normally set from a cross traverse that was designed to have stations close to where the PIs actually fall. The PIs are set from the traverse and checked for distance to the adjacent PI. If the distances are correct the lines are cut out, and the PIs are referenced. Backsights for the references are set out of the construction areas as points on line (POL). Because the PI is not on the curve, it does not have a centerline station number. The PI may have a station number during preliminary reconnaissance of a major transportation route. The PI stations are angle points with deflection angles between straight sections (tangents). After the curves are determined, the entire centerline is restationed. The distance to the PI from the curve/tangent intersection point is found from the tangent formula given in section 9-7.

## 9-7. Curve Formulas

### a. Required parameters

Usually, only two parameters need be specified to totally lay off a curve. If the project drawings contain station numbers, subtract the station number of the PC from the PT station number when using the arc definition. This gives the amount of arc length. The radius or the intersection deflection angle can be used for all other calculations.

$$\text{tangent} \rightarrow T = R \times \tan(\Delta/2)$$

$$\text{chord} \rightarrow ch = 2 \times R \times \sin(\Delta/2)$$

where $\Delta$ = intersection deflection angle or the central angle at the radius for the entire curve.

The relationship of arc to angular measurement can always be used if the radius is known.

$$s = r \times \theta$$

where

$s$ = arc length
$r$ = radius
$\theta$ = angle in radians

Note: If the angle is being used to compute the arc or radius, the units must be in radians. Convert to radians by multiplying degrees by $\pi/180°$.

### b. Relationship between Central Angles and Deflection Angles

The deflection angle measured at the PC between the tangent and the line to the point is one-half the central angle subtended between the PC and the point. The relationship comes directly from the geometry of a circle.

## 9-8. Transition Spirals

The initial factor to determine the transition spiral is the velocity of the vehicle using the structure. Until further guidance becomes available, the formula to be used for determining the minimum length of the spiral will be the highway definition:

$L_S = (1.6 \times V^3)/R_C$

where

$L_s$ = the minimum length of the spiral
$V$ = the design speed (in mph)
$R_c$ = radius of the circular curve

Figure 9-2 is a diagram of a spiral curve used for transition.

$\Delta_S = (L_s \times D_c)/200$

where

$\Delta_S$ = central angle for the spiral
$L_s$ = the length of the spiral used for the spiral design
$D_c$ = degree of curve for the highway curve

The new circular curve will be reduced by

$\Delta_c = \Delta - (2 * \Delta_s)$

where

$\Delta_c$ = new central angle of the circular curve
$\Delta$ = old central angle of the circular curve
$\Delta_s$ = central angle for one spiral

Many procedures exist for computing spiral curves. The method recommended for use by USACE to compute spirals where $\Delta_s$ is less than 15° is

$$X = L_S\left[1 - \frac{\Delta^2}{(5)(2!)} + \frac{\Delta^4}{(9)(4!)} - \frac{\Delta^6}{(13)(6)}\right] \quad 9\text{-}1$$

where $X$ = distance from the tangent to spiral ($T_s$) and PC of the circular curve along the tangent.

$$Y = L_S\left[\frac{\Delta}{3} - \frac{\Delta^3}{(7)(3!)} + \frac{\Delta^5}{(11)(5!)} - \frac{\Delta^7}{(15)(7!)}\right] \quad 9\text{-}2$$

where $Y$ = distance from the tangent to the PC. See Figure 9-2, where

$o = Y - R[1 - \cos(\Delta s)$
$X_O = X - (R)[\sin(\Delta_s)]$
$KG = (R)[\tan(\Delta/2)]$
$FV = (o)[\tan(\Delta/2)]$
$T_s = X_o + KG + FV$
$T_s = X - (R)[\sin(\Delta_s)] + (R + o)(\tan(\Delta_c/2)$

## 9-9. Spiral Stakeout

The point $T_s$ is set from the PI on both tangents. From this point forward, only one side of the spiral will be discussed. The other side is the same. The distance $X$, measured from the $T_s$, is set as a POL. The instrument is moved to the POL, backsighted along the tangent, and the perpendicular is turned to locate the horizontal curve centerline at a distance $Y$ from the tangent line. The deflection angles are computed by the formula

$\delta_S = (l_S^2/L_S^2) * \Delta_S$

The stakeout of a spiral is much the same as a horizontal curve. The arc lengths in the spiral are assumed to be equal to the chords, provided chaining is done between short stations. Fifty-foot stations are common. For an example of a stakeout, assume the $T_s$ falls on station 164+68.21. Assume the $\Delta_s$ was 10°, and $L_s$ is 300.00 ft. To set the first deflection angle with the instrument located at $T_s$, backsighting the PI, subtract out the next even station: 165+00 − 164+68.21 = 31.79 ft (pull 31.79 as the chained distance from $T_s$). The deflection angle is

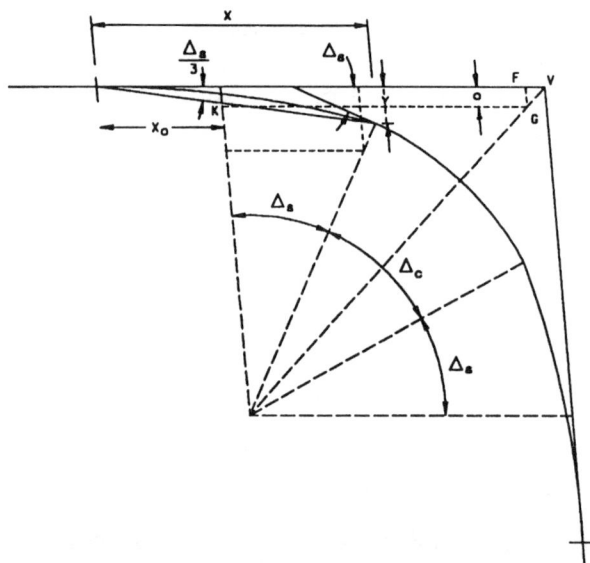

**Figure 9-2. Spiral Curve Diagram**

$\delta_S = 31.79^2/300^2$

$\delta_S = 00° 02' 15''$

A curve (spiral) table is constructed until the last deflection computed before the PC. The deflection from the $T_s$ to the PC is approximately $\Delta_s/3$. No angle in the table should exceed this value.

## 9-10. Vertical Curves

Vertical curves are not typically surveyed to a predetermined design involving topographic surveys. Basically, the same criteria apply for horizontal and vertical curves in preliminary design project phases, which are highly dependent on topographic surveys. Two methods are traditionally used to compute a vertical curve. These are the direct equation method and the tangent offset method. Both methods are discussed. USACE recommends use of the equation method. The tangent offset method offers insight to the calculation of slope and rate of change of slope components to find the elevation on the vertical curve.

## 9-11. Vertical Curve—Tangent Offset Method

Figure 9-3 shows a planview of a straight vertical curve. Vertical curves can be applied to horizontal curves as well. The tangent offsets are computed and algebraically added to the slope elevation computed for the centerline station plus. In Figure 9-3, the tangent offsets reduce all the elevations computed along the tangent to an elevation on the vertical curve. This will not always be the case. Vertical curves have many shapes.

Two definitions are used for the tangent offset method:

1. The parabola is defined as the locus of points equally distant from a point (focus) and a line (directrix).
2. $y = x^2$ is the reduced form of the parabola equation.

An independent parabola is constructed from the known slopes and the length of the curve. Elevations are computed for the point of vertical curve (PVC) and the point of vertical tangent (PVT) (see Figure 9-

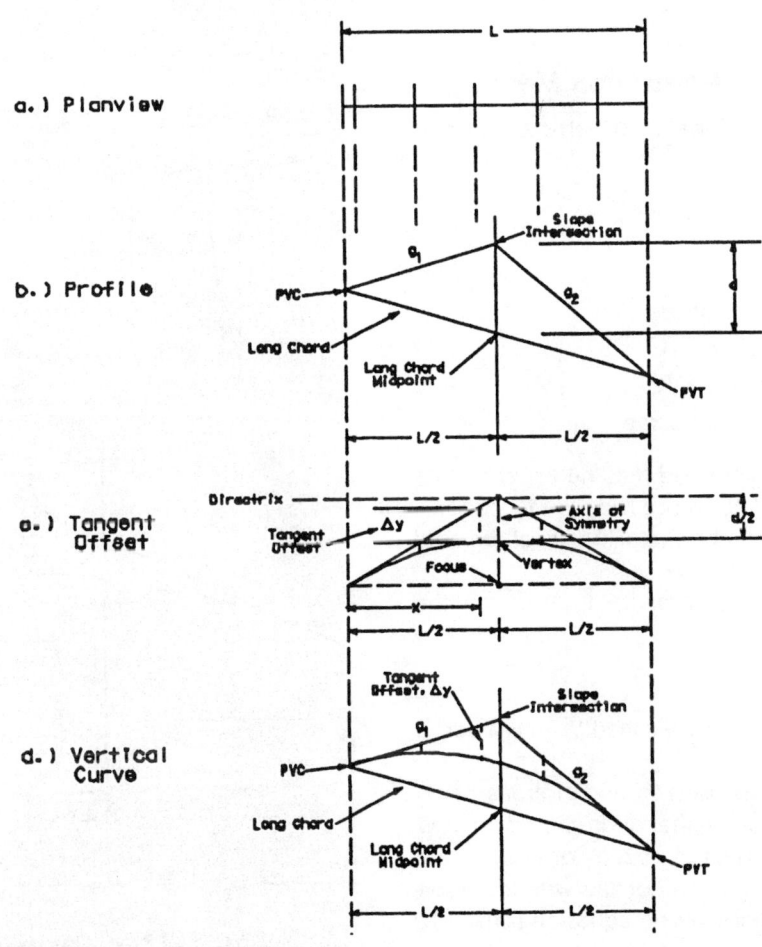

**Figure 9-3. Vertical Curve Geometry**

3, part b). The average of these two elevations is the midpoint of the elevation on the parabola's axis of symmetry. This elevation is substituted for the parabola's focus. The slope intersection is substituted to be the intersection of the axis of symmetry and the directrix (see Figure 9-3, part c). The difference in elevation between the long chord midpoint and the slope intersection is $d$ (see Figure 9-3, part b). A parabola is now constructed with no reference to the actual route alignment until later.

Using first definition, the maximum tangent offset is found along the axis of symmetry between the vertex and the directrix (see Figure 9-3, part c). By definition, the value of this tangent offset is $d/2$.

The vertical distance or tangent offset to any other point on the parabola varies as the square of the horizontal distance from the curve beginning. Both ends of the curve are used if the grades are not equal.

The minimum tangent offset distance is zero on both ends of the parabola. A proportion can be established based on the maximum tangent offset which is stationed at length divided by 2. Combining this with the second definition gives

$$\Delta y / (d/2) = x^2 / (L/2)^2$$

See Figure 9-3, part c.

## 9-12. Vertical Curve—Equation Method

The rate of change of slope in a vertical curve is fixed. This constant is

slope rate equation: $r = (g_2 - g_1)/L$

where

$r$ = rate of change of grade
$g_2$ = grade opposite the PVC
$g_1$ = grade adjacent the PVC
$L$ = length of the vertical curve

Usually, the grades are entered into the equations as percentages, and the lengths are reduced to stations. This makes the rate of change in units of percent grade change per station.

The slope at any point can be found from the Slope Rate Equation as

Slope Equation: $g = rx + g_1$

At the PVC, $x = 0$ and $g = g_1$. At the PVT, $x = L$ and $g = g_2$.

The Slope Equation is used to find stations of no slope. These stations are either high points or low points in the curve. If slopes $g_2$ and $g_1$ are equal, the point of zero slope is on a vertical line with the slope intersection. Otherwise the Slope Equation is used to locate the route station as

$rx + g_1 = 0$

$x = -g_1 / r$

The elevation of any point along the vertical curve can be obtained from the Slope Equation as

$y = (r/2)x^2 + g_1 x +$ elevation of PVC

## 9-13. Vertical Curve Obstructions

Other criteria may impact the design of vertical curves. Obstructions may be the controlling factor in vertical curve design. USACE designs bridges over navigation channels. Shipping commerce must be accommodated in the waterways. A high water elevation in the same datum units as the highway elevations, design shipping clearance, and safety factor provides an obstruction elevation used to compute the length of a vertical curve. Figure 9-4 shows a sketch of a vertical curve and an obstruction elevation ($z$) at a horizontal distance ($s$) along the highway route from the slope intersection of two known grades. The tangent offsets from both ends of the curve to the elevation $z$ are used to compute $L$:

$h_1 / (L/2 + s)^2 = h_2 / (L/2 - s)^2$

$j = i + s \times g_1$

$h_1 = j - z$

$k = i + s \times g_2$ (Note: $g_2$ is negative in the figure.)

$h_2 = k - z$

$$L = 2s \times \left[ \frac{\sqrt{\frac{h_2}{h_1}} + 1}{1 - \sqrt{\frac{h_2}{h_1}}} \right]$$

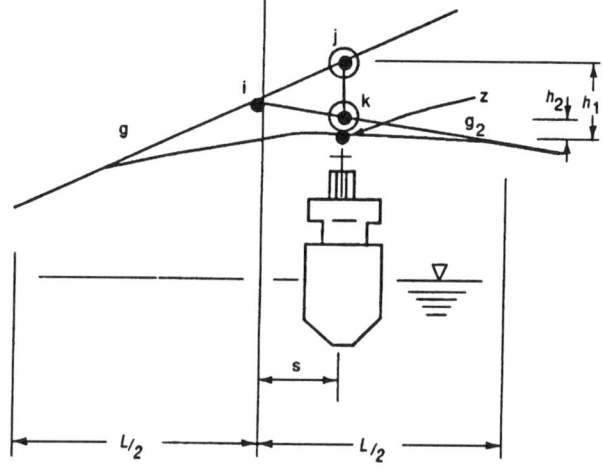

**Figure 9-4. Vertical Curve Obstruction**

# APPENDIX A

# REFERENCES

## A-1. Required Publications

**PL 92-582.** Public Law 92-582 (86 STAT, 1287), "Public Buildings—Selection of Architects and Engineers."

**EFARS 1989.** Engineer Federal Acquisition Regulation Supplement, 1989.

**EP 25-1-1.** Index of Publications, Forms, and Reports Control Symbols.

**ER 405-1-12.** Real Estate Handbook.

**ER 1110-345-710.** Drawings.

**EM 385-1-1.** Safety and Health Requirements Manual.

**EM 1110-1-1000.** Photogrammetric Mapping.

**EM 1110-1-1002.** Survey Markers and Monumentations.

**EM 1110-1-1003.** NAVSTAR Global Positioning System Surveying.

**EM 1110-1-1004.** Deformation Monitoring and Control Surveying.

**EM 1110-1-1807.** Standards Manual for U.S. Army Corps of Engineers Computer-Aided Design and Drafting (CADD) Systems.

**EM 1110-2-1908.** Instrumentation of Earth and Rock-Fill Dams (Earth-Movement and Pressure Measuring Devices).

**EM 1110-2-4300.** Instrumentation for Concrete Structures.

**PARC IL 92-4.** Principal Assistant Responsible for Contracting Instruction Letter 92-4 (PARC IL 92-4), 18 Dec. 1992.

**American Society for Photogrammetry and Remote Sensing, 1989.** "ASPRS Accuracy Standards for Large Scale Maps," *Photogrammetric Engineering and Remote Sensing*, pp. 1068 and 1070.

**Bureau of the Budget, 1947.** *United States National Map Accuracy Standards*, U.S. Bureau of the Budget.

**Davis et al., 1981.** Davis, Raymond E., Foote, Francis S., Anderson, James M., and Mikhail, Edward M., 1981. *Surveying, Theory and Practice*.

**David, Foote, and Kelly, 1966.** Davis, Raymond E., Foote, Francis S., and Kelly, Joe W. (1966). *Surveying, Theory and Practice*.

**Federal Geodetic Control Subcommittee, 1984.** (formerly Federal Geodetic Control Committee) *Standards and Specifications for Geodetic Control Networks*, Rockville, MD.

**Federal Geodetic Control Subcommittee.** *Multipurpose Land Information System Guidebook*.

**Federal Geodetic Control Committee, 1988.** *Geometric Geodetic Accuracy Standards and Specifications for Using GPS Relative Positioning Techniques (Preliminary)*, Rockville, MD. (Reprinted with corrections, 1 Aug. 1989).

**Moffitt and Bouchard, 1987.** Moffitt, Francis H., and Bouchard, Harry (1987). *Surveying* (8th ed.).

# APPENDIX B

# GUIDE SPECIFICATION FOR TOPOGRAPHIC MAPPING SERVICES

## Instructions

### B-1. General

This guide specification is intended for use in preparing Architect-Engineer (A-E) contracts for professional surveying and mapping services. These specifications are applicable to all surveying and mapping contracts used to support U.S. Army Corps of Engineers (USACE) civil works and military design and construction, operations, maintenance, regulatory, and real estate activities. This guide is primarily for use in establishing procedures and specifications obtained through contracts under PL 92-582 (Brooks Act) qualification-based selection procedures and for which unit prices in the contract schedule are negotiated.

### B-2. Coverage

This guide specification contains technical standards and/or references necessary to specify the more common phases of a topographic and planimetric feature detail mapping project performed by conventional methods. Specification of mapping requiring photogrammetric methods are contained in EM 1110-1-1000, *Photogrammetric Mapping*.

### B-3. Applicability

The following kinds of A-E contract actions are supported by these instructions:

- Fixed-price service contracts
- Indefinite delivery type (IDT) contracts
- A multidiscipline surveying and mapping IDT contract in which topographic mapping services are a line item supporting other surveying, mapping, hydrographic, and photogrammetric services
- A work order or delivery order placed against an IDT contract
- Design and design-construct contracts that include incidental surveying and mapping services (including Title II services). Both fixed-price and IDT design contracts are supported by these instructions.

### B-4. Contract Format

The contract format outlined in this guide follows that prescribed in Appendix B of *Principal Assistant Responsible for Contracting Instruction Letter 92-4 (PARC IL 92-4)*, dated 18 December 1992. PARC IL 92-4 incorporates changes to Part 14.201(a)(1) of the 1989 edition of the *Engineer Federal Acquisition Regulation Supplement* (EFARS 1989). The PARC IL 92-4 contract format is designed to support PL 92-582 (SF 252) qualification-based A-E procurement actions.

### B-5. General Guide Use

This guide is primarily intended for field-to-finish topographic and planimetric feature detail survey contracts for large-scale site plan mapping to support engineering design for civil works and military construction projects. The final mapping product with supporting data should be completely suitable for use as a medium to support design and development of contract construction plans and specifications. Specifying field-to-finish implies that all phases of the mapping process, from establishing control, field acquisition, compilation, and delivery of the final compiled product will be performed by the contractor in which the contractor is responsible for complete quality control over all phases of the work.

In adapting this guide to any project, specific requirements will be modified as necessary for the work contemplated. Changes will be made by deletions or insertions within this format. With appropriate adaptation, this guide form may be tailored for direct input in the Standard Army Automated Contracting System (SAACONS). Clauses and/or provisions shown in this guide will be renumbered during SAACONS input.

## B-6. Insertion of Technical Specifications

EM 1110-1-1005, *Topographic Surveying*, should be attached to and made part of any service contract for mapping services. This manual contains specifications and quality control criteria for field-to-finish execution of a mapping project.

Technical specifications for topographic mapping that are specific to the project (including items such as the scope of work, procedural requirements, and accuracy requirements) will be placed under Section C of the SF 252 (Block 10). The prescribed format for developing the technical specifications is contained in this guide. Project-specific technical specifications will not contain contract administrative functions, because these should be placed in more appropriate sections of the contract.

Technical specifications for other survey functions required in a surveying and mapping services contract should be developed from other guide specifications applicable to the discipline(s) required.

Standards and other specifications should be checked for obsolescence and for dates of applicability of amendments and revisions issued subsequent to the publication of this specification. Use EP 25-1-1, *Index of Publications, Forms, and Reports Control Symbols*. Maximum use should be made of existing engineer manuals, technical manuals, and other recognized industry standards and specifications.

## B-7. Alternate Clauses/Provisions or Options

To distinguish between required clauses and optional clauses, required clauses are generally shown in capital letters. Optional or selective clauses, such as would be used in a work order, are generally in lowercase letters.

## B-8. Notes and Comments

General comments and instructions used in this guide are contained in asterisk blocks. These comments and instructions should be removed from the final contract.

## B-9. IDT Contracts and Individual Work Order Assignments

Contract clauses which pertain to IDT contracts, or delivery orders thereto, are generally indicated by notes adjacent to the provision. These clauses should be deleted for fixed-price contracts. In general, sections dealing with IDT contracts are supplemented with appropriate comments pertaining to their use. Work orders against a basic IDT contract may be constructed using the format contained in section C of the contract guide. Clauses in the basic contract should not need to be repeated in work orders. Contract section C is applicable to any type of surveying and mapping service contracting action.

## THE CONTRACT SCHEDULE

## SECTION A

## SOLICITATION/CONTRACT FORM

*********************************************************************************************
**NOTE:** Include here SF 252 in accordance with the instructions in EFARS.
*********************************************************************************************

**SF 252 -- (Block 5): PROJECT TITLE AND LOCATION**

*********************************************************************************************
**NOTE:** Sample title for fixed-price contract
*********************************************************************************************

    TOPOGRAPHIC AND PLANIMETRIC DETAIL SURVEYS IN SUPPORT OF SITE PLAN DEVELOPMENT FOR PRELIMINARY CONCEPT DESIGN OF ENGINEERING INSTALLATION FACILITY, _____ AFB, TEXAS.

    TOPOGRAPHIC MAPPING SERVICES \_\_\_ CHANNEL IMPROVEMENT, \_\_\_\_LOCAL FLOOD PROTECTION PROJECT _____WASHINGTON.

*********************************************************************************************
**NOTE:** Sample title for indefinite delivery type contract.
*********************************************************************************************

    INDEFINITE DELIVERY CONTRACT FOR PROFESSIONAL SURVEYING AND MAPPING AND RELATED SERVICES IN SUPPORT OF VARIOUS *[CIVIL WORKS] [MILITARY CONSTRUCTION] PROJECTS *[IN] [ASSIGNED TO] THE _____ DISTRICT.

*********************************************************************************************
**NOTE:** When other surveying services are also required as part of a broader surveying contract, the clause shown in IL 92-4 shall be used.
*********************************************************************************************

GUIDE SPECIFICATION FOR TOPOGRAPHIC MAPPING SERVICES

## SECTION B
## SERVICES AND PRICES/COSTS

*******************************************************************************

**NOTE:** The fee schedule for topographic mapping and related survey services should be developed in conjunction with the preparation of the independent government estimate (IGE) along with the technical specifications.

*******************************************************************************

| ITEM | DESCRIPTION | QUAN | U/M | U/P | AMOUNT |
|---|---|---|---|---|---|
| 0001 | [Two][Three][Four][__] Man Survey Party; Includes Labor, Travel, Survey Equipment and Materials, Vehicle Cost | | Day | | |
| 0002 | Registered/Licensed Land Surveyor - Office | | Day | | |
| 0003 | Registered/Licensed Land Surveyor - Field | | Day | | |
| 0004 | Supervisory Survey Technician Survey Party Chief | | Day | | |
| 0005 | Survey Technician Survey Instrumentman | | Day | | |
| 0006 | Surveying Aid Survey Rodman/Rodman | | Day | | |
| 0007 | Supervisory Engineering Technician CADD Manager | | M/Hour | | |
| 0008 | Engineering Technician Draftsman/CADD Operator | | M/Hour | | |
| 0009 | Project Manager/Principal | | M/Hour | | |
| 0010 | Computer Charges CADD Processing/Compilation (Operator not included) | | Hour | | |
| 0011 | Compilation Materials/Reproduction | | C/Sht | | |

## SECTION C

## STATEMENT OF WORK

C.1 <u>GENERAL</u>. THE CONTRACTOR, OPERATING AS AN INDEPENDENT CONTRACTOR AND NOT AS AN AGENT OF THE GOVERNMENT, SHALL PROVIDE ALL LABOR, MATERIAL, AND EQUIPMENT NECESSARY TO PERFORM THE PROFESSIONAL SURVEYING AND MAPPING AND *[RELATED SERVICES] *[FROM TIME TO TIME] DURING THE PERIOD OF SERVICE AS STATED IN SECTION D, IN CONNECTION WITH PERFORMANCE OF TOPOGRAPHIC SURVEYS AND THE PREPARATION OF SUCH MAPS AS MAY BE REQUIRED FOR *[ADVANCE PLANNING] [DESIGN] [AND CONSTRUCTION] [or other function] ON [VARIOUS PROJECTS] [specify project(s)]. THE CONTRACTOR SHALL FURNISH THE REQUIRED PERSONNEL, EQUIPMENT, INSTUMENTATION, AND TRANSPORTATION AS NECESSARY TO ACCOMPLISH ALL REQUIRED SERVICES AND FURNISH TO THE GOVERNMENT DETAILED MAPS, PLATS, DIGITAL TERRAIN DATA, UTILITY DETAIL SHEETS, CONTROL DATA FORMS, REPORTS, AND OTHER DATA WITH SUPPORTING MATERIAL DEVELOPED DURING THE FIELD DATA ACQUISITION AND COMPILATION PROCESS. DURING THE PROSECUTION OF THE WORK, THE CONTRACTOR SHALL PROVIDE ADEQUATE PROFESSIONAL SUPERVISION AND QUALITY CONTROL TO ASSURE THE ACCURACY, QUALITY, COMPLETENESS, AND PROGRESS OF THE WORK.

***

**NOTE: The above clause is intended for use in an IDT contract for topographic and planimetric mapping services. It may be used for fixed-price service contracts by deleting appropriate IDT language and adding the specific project survey required. This clause is not repeated on individual delivery orders.**

***

C.2 <u>LOCATION OF WORK</u>.

***

**NOTE: Use the following clause for a fixed-scope contract of individual work order.**

***

    C.2.1. TOPOGRAPHIC MAPPING AND RELATED SURVEYING SERVICES WILL BE PERFORMED AT [_____] *[list project area, installation, etc.]. *[A MAP EXHIBIT DEFINING THE SITE LOCATION AND PROJECT AREA IS ATTACHED AT SECTION G OF THIS CONTRACT.]

***

**NOTE: Use the following when specifying an indefinite delivery contract for topographic mapping services.**

***

    C.2.2 TOPOGRAPHIC MAPPING AND RELATED SURVEYING SERVICES WILL BE PERFORMED IN CONNECTION WITH PROJECTS *[LOCATED IN] [ASSIGNED TO] THE [_____] DISTRICT. *[THE _____] DISTRICT INCLUDES THE GEOGRAPHICAL REGIONS WITHIN *[AND COASTAL WATERS] [AND RIVER SYSTEMS] ADJACENT TO:]

---

*[list states, regions, etc.]

# GUIDE SPECIFICATION FOR TOPOGRAPHIC MAPPING SERVICES

*********************************************************************************************
**NOTE: Note also any local points-of-contact, right-of-entry requirements, clearing restrictions, installation security requirements, etc.**
*********************************************************************************************

C.3 <u>TECHNICAL CRITERIA AND STANDARDS</u>. THE FOLLOWING STANDARDS ARE REFERENCED IN THIS GUIDE. STATE OR LOCAL CODES MAY HAVE PRIORITY.

    C.3.1  USACE EM 1110-1-1005, TOPOGRAPHIC SURVEYING. THIS REFERENCE IS ATTACHED TO AND MADE PART OF THIS CONTRACT. (SEE SECTION G.)

    C.3.2  USACE EM 1110-1-1002, SURVEY MARKERS AND MONUMENTATION. *[THIS REFERENCE IS ATTACHED TO AND MADE PART OF THIS CONTRACT (SEE SECTION G)]

    C.3.3  USACE EM 1110-1-1807, STANDARDS MANUAL FOR USACE COMPUTER-AIDED DESIGN AND DRAFTING (CADD) SYSTEMS.

    C.3.4.  *ASPRS Accuracy Standards for Large-Scale Maps.

    C.3.5.  *[District Drafting Standards, sheet sizes, types, formats, etc.].

    C.3.6.  *[Other applicable references, appendices].

*********************************************************************************************
**Note: List other reference standards that may be applicable to some phase of the work such as other Engineer Manuals or standard criteria documents. Such documents need not be attached to the Contract; if attached, however, reference should be made to their placement in contract Section G.**
*********************************************************************************************

C.4 <u>WORK TO BE PERFORMED</u>. PROFESSIONAL SURVEYING AND MAPPING AND RELATED SERVICES TO BE PERFORMED UNDER THIS CONTRACT ARE DEFINED BELOW. UNLESS OTHERWISE INDICATED IN THIS CONTRACT *[OR IN DELIVERY ORDERS THERETO], EACH REQUIRED SERVICE SHALL INCLUDE FIELD-TO-FINISH EFFORT. ALL MAPPING WORK WILL BE PERFORMED USING APPROPRIATE INSTRUMENTATION AND PROCEDURES FOR ESTABLISHING CONTROL, FIELD DATA ACQUISITION, AND COMPILATION IN ACCORDANCE WITH THE FUNCTIONAL ACCURACY REQUIREMENTS TO INCLUDE ALL QUALITY CONTROL ASSOCIATED WITH THESE FUNCTIONS. THE WORK WILL BE ACCOMPLISHED IN STRICT ACCORDANCE WITH SURVEYING AND MAPPING CRITERIA CONTAINED IN THE TECHNICAL REFERENCES (PARAGRAPH C.3 ABOVE), EXCEPT AS MODIFIED OR AMPLIFIED HEREIN.

*********************************************************************************************
**NOTE: The following clauses in this section of the guide may be used for either fixed-price surveying and mapping contracts, IDT work orders under an IDT contract, or IDT contracts where surveying and mapping services are part of a schedule of various survey disciplines.**
*********************************************************************************************

    C.4.1. <u>PURPOSE OF WORK</u>. THE WORK TO BE PERFORMED UNDER THIS CONTRACT IS TO BE USED AS BASIC SITE PLAN MAPPING INFORMATION TO SUPPORT *[BE INCORPORATED INTO] [INSTALLATION/ BASE COMPREHENSIVE MASTER PLANNING] [ENGINEERING DESIGN] [CONSTRUCTION] [OPERATION]

[MAINTENANCE] [REAL ESTATE] [REGULATORY] [HAZARDOUS AND TOXIC WASTE SITE _____];
INCLUDING ALL RELATED ACTIVITIES.

***

NOTE: A description of the functional purpose of the mapping product should be stated in order for the contractor to focus his efforts and quality control toward the more critical aspects of the project. The above clause should fully define the intended use of the mapping product to be furnished by the contractor.

***

C.4.2. <u>GENERAL SURVEYING AND MAPPING REQUIREMENTS</u>. TOPOGRAPHIC AND PLANIMETRIC FEATURE DETAIL MAPS SHALL BE COMPILED AT A TARGET SCALE OF 1 IN. = [____] FT FOR THE SITE DELINEATED ON EXHIBIT ATTACHED AT SECTION G. THE MAPPING AND/OR RELATED DIGITAL PRODUCTS SHALL MEET OR EXCEED USACE (ASPRS) CLASS *[____] ACCURACY STANDARDS AS SPECIFIED IN EM 1110-1-1005. PLANIMETRIC FEATURE DETAIL WILL BE COMPILED IN ACCORDANCE WITH THE HORIZONTAL ACCURACY STANDARDS SET FOR THIS CLASS. CONTOURS SHALL BE DEVELOPED AT [____]-FT INTERVALS IN ACCORDANCE WITH THE VERTICAL ACCURACY STANDARDS SET FOR THIS CLASS. FEATURE AND TERRAIN DATA SHALL BE DELIVERED IN *[HARD COPY AND] DIGITAL FORMAT.

***

NOTE: The above clause should be used for fixed-scope contracts of IDT contract work orders to give an overview of the general mapping effort. Technical requirements will be described in subsequent paragraphs.

Note that the final map compilation target scale and ASPRS Accuracy Class/Standard is defined upfront in the scope of work.

IDT contracts and work orders: Since specific project scopes are indefinite at the time a basic contract is prepared, only general technical criteria and standards can be outlined. Project of site-specific criteria will be contained in each delivery order along with any deviations from the technical standards identified in the basic IDT contract. The clauses contained within the remainder of the contract are used to develop general requirements for a basic IDT contract. Subsequent delivery orders will reference these clauses, adding project-specific work requirements as required. Delivery order formats should follow the outline established for the basic IDT contract.

***

C.4.3. <u>FIELD PROCEDURES AND REQUIREMENTS</u>. APPROPRIATE INSTRUMENTATION AND PROCEDURES, CONSISTENT WITH ACCEPTED PROFESSIONAL SURVEYING AND MAPPING INDUSTRY STANDARDS AND PRACTICE, SHALL BE SELECTED TO ACHIEVE THE ACCURACY STANDARDS REQUIRED. THE CONTRACTOR SHALL FIELD A FULLY EQUIPPED SURVEY CREW(S), CONSISTING OF PROFESSIONAL SURVEY PERSONNEL, EXPERIENCED IN PERFORMING THE REQUIRED SURVEYS AND CAPABLE OF COMPLETING THE WORK WITHIN ALLOTTED SCHEDULES. ALL FIELD OBSERVATIONAL DATA REQUIRED TO SET AND ESTABLISH PROJECT CONTROL SHALL BE RECORDED IN STANDARD PERMANENT BOUND FIELD BOOKS WHICH WILL SUBSEQUENTLY BE DELIVERED TO THE GOVERNMENT. ALL SURVEY WORK SHALL BE PERFORMED UNDER ADEQUATE SUPERVISION AND QUALITY CONTROL MEASURES. *[ALL SURVEY WORK, INCLUDING OFFICE COMPUTATIONS AND ADJUSTMENTS, IS SUBJECT TO GOVERNMENT REVIEW AND APPROVAL FOR CONFORMANCE WITH PRESCRIBED ACCURACY STANDARDS. DEFICIENCIES WILL BE RECOGNIZED AND STEPS TO INITIATE

CORRECTIVE ACTIONS SHALL BE TAKEN AS REQUIRED]. *[THE CONTRACTOR SHALL ALLOW DIRECT CONTACT WITH RESPONSIBLE-IN-CHARGE PERSONNEL FOR EACH PHASE OF THE WORK FOR PURPOSES OF PROGRESS ESTIMATES AND COMPLIANCE WITH THE CONTRACT REQUIREMENTS].

C.4.3.1. HORIZONTAL CONTROL SHALL REFERENCE EXISTING PROJECT AREA CONTROL. CONTROLLING POINTS SHALL BE OCCUPIED AS A STATION WITHIN A CLOSED TRAVERSE THAT WILL MEET OR EXCEED *[THIRD][___]- ORDER, *[CLASS*[I][_] RELATIVE ACCURACY CLASSIFICATION *[OR 1 PART IN 10,000] [____] AS ESTABLISHED FOR ASPRS CLASS *[__] MAPPING STANDARDS. THE TRAVERSE SHALL INITIATE AND CLOSE UPON ACCEPTABLE CONTROL MONUMENTATION USED TO ESTABLISH THE EXISTING PROJECT GRID SYSTEM. ALL GRID COORDINATES SHOWN ON THE MAP PRODUCTS SHALL BE EXPRESSED IN OR CONVERTED TO, *[US SURVEY FEET] [INTERNATIONAL FEET] [METERS]. COORDINATES SHALL BE REFERENCED TO THE LOCAL *[SPCS 27] [SPCS 83] [UTM ZONE].

C.4.3.2. VERTICAL CONTROL SHALL BE REFERENCED TO *[NGVD 29][NAVD 88]. CONTROLLING POINTS SHALL BE ESTABLISHED WITHIN A CLOSED LEVEL LOOP THAT WILL MEET OR EXCEED *[THIRD] [_____]-ORDER, ACCURACY STANDARDS AS ESTABLISHED FOR ASPRS CLASS *[___] MAPPING STANDARDS. ELEVATIONS SHALL ORIGINATE AND CLOSE ON ACCEPTABLE BENCHMARKS IN THE PROJECT AREA. UNLESS OTHERWISE INDICATED, INITIATING AND CLOSING THE LEVEL LOOP ON THE SAME BENCHMARK SHALL NOT CONSTITUTE AN ACCEPTABLE CONTROL CIRCUIT.

***

**NOTE: Few USACE surveying and mapping projects require relative accuracy classifications in excess of Third Order, Class I, 1:10,000 for horizontal control and Third Order 1:5,000 for vertical control. Although instrumentation, conventional and GPS, are capable of achieving higher accuracy requirements, specifying higher levels of accuracy may adversely impact the overall project cost and should be thoroughly justified relative to the required mapping accuracies and other factors.**

***

C.4.3.3. EXISTING PROJECT/NETWORK CONTROL. A TABULATION AND/OR DESCRIPTION OF EXISTING PROJECT/NETWORK CONTROL POINTS *[IS SHOWN BELOW] [IS SHOWN IN ATTACHMENT] [WILL BE PROVIDED IN THE DELIVERY ORDER]. THE SOURCE AGENCY, COORDINATES, DATUM, AND ESTIMATED ACCURACY OF EACH POINT ARE INDICATED. PRIOR TO USING ANY CONTROL POINTS, THE MONUMENTS SHOULD BE CHECKED TO ENSURE THAT THEY HAVE NOT BEEN MOVED OR DISTURBED.

***

**NOTE: List existing control station(s) or, alternately, refer to a map exhibit, tabulation attachment, and/or descriptions that would be included in an attachment.**

***

a. *The contractor shall perform the necessary surveys to connect existing project control to assure that such control has sufficient relative accuracy to adequately control the overall project. Should these surveys indicate deficiencies in the existing control, the contractor shall advise the Contracting Officer *[or Contracting Officer Representative]. The contractor shall furnish the appropriate data indicating a deficiency. If the Contracting Officer *[or Contracting Officer Representative] deems it necessary to perform resurveys of the existing network, appropriate modification may be made to the contract.

b. *Suitable control monumentation shall be set as required to adequately control construction phases. All stations shall be monumentated in accordance with EM 1110-1-1002, Survey Markers and Monumentation. Monumentation for this project shall be Type *[___] for horizontal and Type *[___] for vertical, per EM 1110-1-1002 criteria. *[Monumentation shall be defined to include the required reference marks and azimuth marks required by EM 1110-1-1002.]

*******************************************************************************
**NOTE: Deviations from EM 1110-1-1002 should be indicated as required. USACE project control rarely requires supplemental reference/azimuth marks - the optional specification clauses below should be tailored accordingly.**
*******************************************************************************

c. *At each station, angle and distance measurements shall be made between a network station and reference/azimuth marks established in accordance with the requirements set forth in EM 1110-1-1002. All observations shall be recorded in a standard bound field book.

(1) *For reference marks, two (2) directional positions are required (reject limit ± 10-second arc) and with steel taping performed to the nearest ± 0.01 ft.

(2) *Four directional positions are required for azimuth marks. The reject limit for a 1-second theodolite is ± 5 seconds. Azimuth mark landmarks shall be easily defined/described natural features or structures of sufficient distance to maintain a *[  ]-second angular accuracy. *[   -order astronomic azimuths shall be observed to azimuth marks.]

(3) *A compass reading shall be taken at each station to reference monuments and azimuth marks.

C.4.3.4. <u>STATION DESCRIPTION AND RECOVERY REQUIREMENTS</u>.

a. *Station descriptions and/or recovery notes shall be written in accordance with the instructions contained in EM 1110-1-1002. *[Form [____]shall be used for these descriptions.] Descriptions shall be *[written] [typed].

b. *Descriptions *[are] [are not] required for *[existing] [and/or newly established] stations.

c. *Recovery notes *[are] [are not] required for existing stations.

d. *A project control sketch *[is] [is not] required.

C.4.4. <u>FIELD CLASSIFICATION AND MAP CHECK SURVEYS</u>. FIELD CLASSIFICATION, INSPECTION, AND/OR CHECK/MAP ACCURACY TEST SURVEYS *[WILL] [WILL NOT] BE PERFORMED *[ON THIS PROJECT].

*******************************************************************************
**NOTE: Tests for compliance of a map sheet are optional. Check points are based on "well-defined points" established in a manner agreed upon by the contracting parties. Criteria for testing for map accuracy compliance are defined in the ASPRS Accuracy Standards for Large-Scale Maps.**
*******************************************************************************

C.5. <u>MAP COMPILATION, DRAFTING, AND CADD SPECIFICATIONS</u>.

C.5.1. <u>MAP COMPILATION SCALE</u>. THE CONTRACTOR SHALL FURNISH *[REPRODUCIBLE] FINISHED MAPS AT A SCALE OF 1 IN. = *[   ] FT. THE MAP MEDIA SHALL BE COMPUTER-GENERATED PLOTS ON *[PAPER] [HIGH-GRADE, STABLE BASE MYLAR NOT LESS THAN *[____] IN. IN THICKNESS] *[E] [_] - SIZE SHEETS].

## SECTION F

## CONTRACT CLAUSES

*************************************************************************************************
**NOTE:** See instructions in Appendix B of PARC IL 92-4.
*************************************************************************************************

## SECTION G

## LIST OF ATTACHMENTS

G.1  U.S. ARMY CORPS OF ENGINEERS EM 1110-1-1005, TOPOGRAPHIC SURVEYING.  THIS REFERENCE IS ATTACHED TO AND MADE PART OF THIS CONTRACT.

*************************************************************************************************
**NOTE:** List any other attachments called for in contract Section C or in other contract sections. This may include such items as:

   a. Marked-up exhibits, project sketches/drawings.
   b. Station/Monument descriptions or Recovery Notes.
   c. Drafting Standards.
   d. CADD Standards.
*************************************************************************************************

## SECTION H

## REPRESENTATIONS, CERTIFICATIONS, AND OTHER STATEMENTS OF OFFERERS

## SECTION I

## INSTRUCTIONS, CONDITIONS, AND NOTICES TO OFFERERS

*************************************************************************************************
**NOTE:** See PARC IL 92-4 for guidance in preparing these clauses/provisions.
*************************************************************************************************

# APPENDIX C

# AUTOMATED TOPOGRAPHIC SURVEY DATA COLLECTOR EQUIPMENT: INVENTORY OF USACE INSTRUMENTATION AND SOFTWARE

| District | Total Station | Data Collector | Software | Portable |
|---|---|---|---|---|
| Alaska | Wild T2000 (1) <br> Wild T1000 (1) | GRE 3 (1) | Wildsoft | Zenith (1) |
| Albuquerque | Geodimeter 140 (1) | None | None | None |
| Baltimore | None | None | None | None |
| Buffalo | Topcon GTS 3B (3) | FC-1 (1) | None | None |
| HEC | None | None | None | None |
| Chicago | None | None | Civilsoft | None |
| CERL | Wild T2000 (1) | GRE 3 (1) | Maine Surveyors (FOG) | None |
| Detroit | Geodimeter 440 (1) <br> Wild T2000 (1) <br> Wild T2002 (1) <br> Topcon GTS 3B (1) <br> Topcon GTS 2 (1) <br> Nikon DTMS (1) | Geodat 126 (1) <br> GRE 3 (1) <br> GRE 4 (1) <br><br><br> DR-1 (1) | Geodimeter <br> Wildsoft <br> Wildsoft <br> Pacsoft | Compaq 386 (1) |
| Detroit (GH) | Wild T2000 (1) <br> Wild T2002 (1) | GRE 3 (1) <br> GRE 4N (1) | Wildsoft | None |
| Detroit (Duluth) | Topcon GTS 3B (1) | None | Wildsoft | None |
| TEC | None | None | None | Compaq 386 (1) |
| Fort Worth | Lietz SetS3 (1) | SDR 24 (1) | Lietz SDRMAP | None |
| Galveston | HD 3800 | None | None | None |
| Galveston | Lietz Set2 (1) | SDR 2 (1) | Lietz SDRMAP | None |
| Huntington | Wild TC 1600 (1) <br> Nikon DTMS5 (1) | GRE 4 (1) <br> DR-2 (1) | CEORH ED-S | None |
| Huntsville | None | None | None | None |

# AUTOMATED TOPOGRAPHIC SURVEY DATA COLLECTOR EQUIPMENT: INVENTORY

| District | Total Station | Data Collector | Software | Portable |
|---|---|---|---|---|
| Jacksonville | Wild T2000 | GRE 3 (1) | None | None |
| Kansas City | None | None | None | None |
| Little Rock (GD) | Lietz Set2 (1)<br>HP 3810 A (1) | SDR 2 (1)<br>HP71B (2) | Lietz Cogo Plus<br>ETI | Compaq 386 (2) |
| Little Rock (PB) | None | None | None | None |
| Los Angeles | None | None | None | Zenith (1) |
| Louisville | Geodimeter AD (4) | Geodat 124 (4) | Geodimeter | Zenith (4) |
| Memphis | Wild T2000 (2) | GRE 3 (2) | Pacsoft | Compaq (2) |
| Mobile | Topcon GTS-30 (1) | None | None | Grid 1139 (2) |
| Nashville | Wild TC 2000 (1)<br>Topcon GTS-2 (2) | GRE 4 (1)<br>BTI | ETI | None |
| New Orleans | Lietz Set2 (1) | SDR 24 (1) | SDRMAP | None |
| New England | Topcon ET-1 (1)<br>Topcon ET-2 (2) | Husky Hunter | | None |
| Norfolk | None | None | None | None |
| North Atlantic | None | None | None | None |
| Omaha | Wild T2000 (2)<br>Wild TG2000 (1) | GRE-3 (1)<br>GRE-# (2) | Pacsoft<br>ETI<br>ESP-200 | Zenith (3) |
| Pacific Ocean | None | None | None | None |
| Philadephia | Topcon GTS-2 (1) | None | None | None |
| Pittsburgh | Topcon GTS-2 (1)<br>Topcon GTS-3 (2)<br>Nikon NO-5 (1) | FC-1 (1) | ETI | None |
| Portland | T2000 (1)<br>T2002 (1)<br>Geodimeter 140 (1) | GRE 4 (2)<br>Geodat 122 (2) | Geodimeter | Zenith (1) |
| Rock Island | Wild T2000 | GRE 4 (1) | Wildsoft | Zenith 286 (1) |
| Sacramento | Geodimeter 136 (3)<br>Geodimeter 444 (3) | Geodat 126 (2) | Geodimeter<br>Wild | Zenith (1) |
| Savannah | Geodimeter 140 (4)<br>Topcon GTS-36 (1) | HP 41 CX<br>HP 41 CX | Mount Gilmore | None |
| South Pacific | None | None | None | None |
| St. Louis | HP 3820A (1)<br>Geodimeter 440 LR | Maptech (1)<br>Geodat 400 (1) | Maptech<br>MapCAD | HP 71B (2) |
| St. Paul | Wild T2000 (1) | GRE-3 (1) | Wildsoft | Compaq 386 (1) |
| Seattle Hydro | Geodimeter 140 (1)<br>Geodimeter 449 (1) | Geodat 126 (2) | None | None |
| Seattle | Wild T2000 | GRE-3 (1) | Wildsoft | Zenith (1) |
| Vicksburg | Lietz Set2 (2) | SOR 2 (1) | SDRMAP | None |
| WES | None | None | None | None |

# APPENDIX D

# COORDINATE GEOMETRY SOFTWARE

## D-1. General

COGO (**CO**ordinate **GeO**metry) was initially developed by Charles L. Miller of M.I.T. in 1959. Since then, many improvements have been made, but the basic concept and vocabulary have remained the same. COGO is a problem-oriented system that enables the user with limited computer experience to solve common surveying problems. The language is based on familiar surveying terminology such as, Azimuth, Inverse, Bearing, etc. This terminology is used to define the problem and generate a solution. COGO may be used to solve problems such as curve alignments, point offsets, distance and direction between two points, intersections, etc.

The basis of the system is a series of commands used to manipulate or compute points defined by a point number, x-coordinate, and y-coordinate on a plane surface. These points are stored in what is referred to as the "coordinate table" and may be recalled by their point number in future computations.

The mathematics used for the computations described in this appendix are beyond the scope of this manual. There are many books published that describe the mathematical procedures in detail.

Many COGO packages are on the market today. Several are available to the U.S. Army Corps of Engineers (USACE) free of charge. Among these are

- U0002, by Waterways Experiment Station
- MCOGO, by Simple Survey Software Inc.
- BLM-COGO, produced by the Bureau of Land Management

All of these are available from the Topographic Engineering Center.

## D-2. Requirements

Requirements for COGO are as follows:
- The ability to utilize a *combined scale factor* in its computations. This will allow the user to calculate the ground distances when staking out a job, or reduce the measured distances to the reference vertical datum, and correct for the scale factor when the survey is to be tied to the State Plane Coordinate System (SPCS).
- The ability to rotate and scale (transform) the survey points to fit existing control. When the surveyor uses field coordinates to perform the survey job, the survey can be transformed onto the SPCS by defining two points with their SPCS coordinates.
- Compass traverse adjustment is sufficient for the majority of traverses established by the USACE. Ability to perform a least-squares adjustment can be more advantageous.
- Must have the ability to work in bearings, north azimuth, or south azimuth.
- Allow the export of the coordinate table to an ASCII file.
- Allow the import of points from an ASCII file. Typically, this file will be the ASCII coordinate file produced by total stations and their associated software.

## D-3. Functions

COGO functions can be grouped into many categories.

### a. Forward Computation Commands

Used to calculate the coordinates for a point, given the coordinates of a known point and the distance and direction to the unknown point.

1. LOCATE/AZIMUTH: Computes a point given an azimuth and direction from a known point
2. LOCATE/BEARING: Computes a point given a bearing and direction from a known point
3. LOCATE/ANGLE: Computes a point given a backsight point, angle, and distance
4. LOCATE/LINE: Computes a point on tangent (POT) given tangent end points and a distance
5. LOCATE/DEFLECTION: Computes a point given a backsight, deflection angle, and a distance

### b. Inverse Computation Commands

Used to compute the distance and direction between two known points. Both the ground and grid distances should be given as output.

1. INVERSE/AZIMUTH: Computes the distance and azimuth between two known points
2. INVERSE/BEARING: Computes the distance and azimuth between two known points
3. TANGENT/OFFSET: Computes the distance offline and the distance downline given a known point and the ends of a known tangent

### c. Intersection Commands

Used to calculate the coordinates of an unoccupied point as the intersection of two vectors of defined direction and/or distance from two known points.

1. LINE/LINE INT: Computes the coordinates of the point of intersection of two lines whose end points are known
2. RANGE/RANGE INT: Computes the coordinates of the intersection of two arcs with known radii and centers. Two answers are possible, so the user must define the desired intersection.
3. RANGE/AZIMUTH INT: Computes the coordinates of the intersection of a defined vector and an arc. Two answers are possible, so the user must define the desired intersection.
4. AZIMUTH/AZIMUTH INT: Computes the coordinates of the intersection of vectors with known direction.
5. FORESECTION: This is an Azimuth/Azimuth intersection, measured by turning angles from two known points.

### d. Curve Commands

Allow the user to define curve parameters to use defined alignment in computations.

1. ALIGNMENT: Given measured curve parameters, computes components of a curve such as
   - arc length
   - long chord
   - radius
   - degree of curve
   - tangent length
   - center point coordinates
   - external distance
   - mid-ordinate
   - central angle
2. STATION/OFFSET: Computes the coordinates of an unknown point, given a station and offset along the curve. The reverse function is also available to compute the station and offset of a known point relative to the curve alignment.

### e. Area Computation Commands

Calculate the area of polygons and curve segments. The COGO package should calculate the area based on ground distances, not the reduced distances. This is done by multiplying the computed area by the combined scale factor squared.

# APPENDIX E

# SAMPLE SCOPE OF WORK

**DETAILED SPECIFICATIONS**

**WORK ORDER NO. -----**

1. PROJECT: Survey III Mapping Project

2. LOCATION: Huntsville, Alabama

3. GENERAL: Perform field topographic, planimetric, and utility surveys, office computations, and 3D digital mapping for use in developing "Plans and Specifications" as detailed in this scope of work.

4. SPECIFIC REQUIREMENTS

   a. Horizontal control for the project shall comply with Corps of Engineers Third-Order standards as outlined in EM 1110-1-1005.

   b. Vertical control shall be Fourth-Order per EM 1110-1-1005.

   c. Using control (NAD 83 and NGVD 29) provided by the Government, the Contractor shall lay out horizontal and vertical control in project areas. Control points shall be semi-permanent (re-bar w/cap) and set in a manner that they can be used for layout during construction. From said control points, the Contractor shall acquire field topographic (cross-sections or random topo) and planimetric information (buildings, roads, parking areas, sidewalks, fence lines, structures, drainage, etc.) to be used for 40 scale mapping with a one (1) foot contour interval. Density of field elevations shall support 1" = 40' mapping and shall be provided as necessary to show all breaks in grade or changes in terrain. Also the Contractor shall locate and tie individual trees (size and species) in the project area. All elevations shall be taken to the hundredth of a foot. All horizontal and vertical data will be collected with an electronic data collector using the Government-furnished data collection codes. The Contractor shall also simultaneously record data in the field book. This will include setups, backsights, measure ups, shot numbers, and shot descriptions. Angles and distances shall be recorded for every + 20 shots. The Contractor shall furnish the Government the field files (collection and the edited and compiled field file) along with the final coordinate file for all work. Vertical control for utilities shall be taken with a total station instrument, with measure downs for the invert elevations. All vertical control for utilities shall be recorded in field along with any sketches required. Utility information is required for the following:

   (1) Water: Locate all valves, standpipes, regulators, etc. Locate all fire hydrants. Provide an elevation on top of valve case and top of valve. Provide size of pipe and distance above ground for standpipes.

   (2) Sanitary Sewer: Locate all manholes and provide top of rim elevation along with an invert elevation of all pipes connected to the manhole. Identify type, size, and direction of each pipe.

   (3) Storm Drainage: Locate manholes and all other storm drainage structures such as culverts, headwalls, catch basins, and clean-outs. Provide top of manhole or top of catch basin elevation along with an invert elevation of all pipes connected to a manhole or catch basin and bottom elevation. Identify type, size, and direction of each pipe. Provide type, size, and invert elevation for all culverts.

(4) Electrical: Locate all power poles, guy wires, vaults, manholes, meters, transformers, electrical boxes, and substations. Obtain type and height of poles, number and size of transformers, number of crossarms, number of wires (electrical and communication), and direction and low wire elevation at each pole. Provide top of rim or top of vault elevation, top of wire or conduit elevation, and direction and bottom elevation of manholes and vaults. Provide size for all electrical vaults and boxes.

(5) Gas: Locate all valves, meters, and gas line markers. Provide elevation on top of valve case and on top of valve.

(6) Telephone: Locate all poles, manholes, boxes, etc. Provide top of rim elevation, top of wire or conduit elevation, and direction and bottom of manhole elevation. Obtain type and height of poles, number of crossarms, number of wires, and low wire elevation at each pole.

(7) Street Light: Locate all poles and provide type and height of poles. Identify number and type of lights on poles. If connected by wires, show direction and low wire elevation.

(8) Heating: Locate all steam manholes and vaults, filler pipes, underground fuel tanks, etc. Provide top of rim or top of vault elevation, top of pipe elevation, and direction and bottom of pit elevation. Provide size of vault and all pipe sizes within manhole or vault.

(9) Fire Alarm: Locate any fire alarm systems (box with number), telephones (box with number), etc. in project.

d. All computations are to be arranged in a sequential and understandable order, with notes when appropriate so a review can be made with minimum reconstruction. The Contractor shall furnish the Government computer output of unadjusted bearings, azimuths, distances, and coordinates of all traverse points. The error of closure, both azimuth and positions, shall be shown. Final data will be adjusted by the compass method and will show the adjusted bearings, distances, and coordinates of all points surveyed. The Contractor shall provide a final list of coordinates for all points. The Contractor shall use the combined grid factor for all work. All level lines shall be reduced and adjusted in accordance with accepted procedures and practices. All computations shall be fastened into an 8.5" × 11" folder separated and labeled to indicate various facets of work (horizontal, vertical).

e. Field note books standards:

(1) Field books shall be neat, legible, and sequential. They also will show names of crew members and date at the beginning of eachday. Black ink shall be used.

(2) Each field book shall have an index. The serial number and type of instruments used will be shown on this page.

(3) There is to be a maximum of one (1) horizontal setup per page.

f. Target computer system: The Contractor shall provide interactive graphic and nongraphic data files that are fully operational on an Intergraph computer system running MicroStation software, version 4.0 or better. The files shall be created using Government-furnished seed file to ensure compatibility with mapping procedures and standards.

g. Utility information. All utilities that are field tied shall be merged into the Government-furnished 1" = 40' topographic database. This includes showing manholes, valves, power poles, etc., and connecting lines. Also the attribute information (text) for each utility shall be placed in the data file. This can include but not be limited to top of rim elevations, invert elevations, pipe size, direction, top of valve elevation, etc. (See Government-furnished example.) All horizontal and vertical control established for ties shall be shown as a symbol with annotation. (Also, see Appendix A for breakdown of level assignments, level symbology, and text size.)

h. Map symbols. All symbols shall conform with Government-furnished cell file (CIVSUR.CEL). (See Appendix B for a complete breakdown of cells.)

i. Global origin. The Contractor shall use the standard global origin of zero $X$- and $Y$-coordinates at the lower left corner of the $X$-$Y$ plane.

j. Views. Only views one (1) and five (5) will be active. All locks will be off except keypoint snap, and all displays will be on except text nodes and grid.

k. Text/Font. Most map features constitute either graphics or text and are on separate levels. However, in some cases, text will be placed on the same level as the graphics. Examples of this would be the "S" embedded in the line for sanitary sewer or the "W" in the line for water. Font 24 shall be used for utility descriptions and font 127 for all remaining text. (See Appendix A for breakdown of level assignments, level symbology, and text size.)

## 5. SPECIAL REQUIREMENTS

a. There shall be no cutting of trees, and brush cutting shall be kept to a minimum.

b. Excessive marking with paint, flagging, etc. will be avoided.

c. The Contractor shall comply with all applicable safety regulations of the current U.S. Army Corps of Engineers Safety and Health requirements manual EM 385-1-1, and shall acquaint himself and his personnel with the safety requirements governing the area in which the work is being done.

## 6. MATERIAL TO BE FURNISHED TO CONTRACTOR

a. Utility maps (1" = 40') as required for areas of work:
(1) storm drainage
(2) sanitary sewer
(3) water
(4) electrical
(5) street lighting
(6) telephone
(7) gas
(8) fire alarm
(9) heating

b. Control listing and map

c. Collection Point Codes, font library (FONTLIB.NPS), cell file (CIVSUR.CEL), and seed file (SEED.DGN)

d. Field book example

e. Final product example (1" = 40' plot)

f. Appendix A, Level Assignments and Symbology

## 7. REVIEW/SUBMITTAL

a. Initial submittal: The Contractor shall provide the Government one (1) completed Intergraph design file and hard copy at 1" = 40' for review to assure compliance with project specifications. The Government reserves a period of five (5) calendar days to comment on this submittal.

b. Pre-final submittal: The Contractor shall generate 3D graphic files of utility data (Intergraph design files) and a 1" = 40' plot of all files for all areas. Digital data shall be supplied on a 5.25-in. ,1.2-MB floppy disk, 3.5-in., 1.44-MB floppy disk, or a 5.25-in., 44-MB removable cartridge. The Government reserves a period of ten (10) calendar days to comment on Contractor's work.

c. Final submittal: The final submittal shall contain all the revisions required as a result of the Government's pre-final review. The final submittal shall consist of
(1) Intergraph design files on 3.5-in. or 5.25-in. floppy disk or cartridge for each file
(2) 1" = 40' plots of individual data files
(3) all items in paragraph 6 above
(4) all computations (in folder)
(5) all field books (reduced and checked)
(6) floppy disks of all raw field data and final coordinate data

8. REPORTS: The Contractor shall submit monthly progress status reports during the duration of the project.

9. SCHEDULE AND DELIVERY: The submission schedule shall commence on the day notice-to-proceed is issued and will run consecutively for the number of days shown in Appendix B. All submittals shall be accompanied by a letter of transmittal.

Enclosures:

Appendix A

Appendix B

[Appendices to be included in an actual scope of work.]

# APPENDIX F

# GLOSSARY

**A-E:** Architect-Engineer
**AM/FM:** Automated Mapping/Facilities Management
**ANSI:** American National Standards Institute
**ASCE:** American Society of Civil Engineers
**ASCII:** American Standard Code for Information Interface
**ASPRS:** American Society of Photogrammetry and Remote Sensing
**CADD:** computer-aided drafting and design
**CMAS:** Circular Map Accuracy Standards
**COGO:** coordinate geometry
**DEM:** Digital Elevation Model
**DGN:** Intergraph design
**2DRMS:** two standard deviations root-mean-square
**DTM:** digital terrain model
**EDC:** electronic distance collection
**EDM:** electronic distance measurement
**EFARS:** Engineer Federal Acquisition Regulation Supplement
**FGCC:** Federal Geodetic Control Committee
**FGCS:** Federal Geodetic Control Subcommittee
**GIS:** Geographic Information System
**GPS:** Global Positioning System
**GRS 80:** Geodetic Reference System of 1980
**HARN:** High Accuracy Regional Network
**HI:** height of instrument
**IDT:** indefinite delivery type
**IGES:** Initial Graphic Exchange Specification
**LIS:** Land Information System
**NAD 27:** North American Datum of 1927
**NAD 83:** North American Datum of 1983
**NAVD 88:** North American Vertical Datum of 1988
**NGRS:** National Geodetic Reference System
**NGS:** National Geodetic Survey
**NGVD 29:** National Geodetic Vertical Datum of 1929
**NMAS:** National Map Accuracy Standards System
**NOS:** National Ocean Service
**OMA:** Operations and Maintenance Army
**OMAF:** Operations and Maintenance Air Force
**OMB:** Office of Management and Budget
**PC:** point of curvature
**PI:** point of intersection
**POL:** points on line
**PT:** point of tangency
**RMS:** root-mean-square
**RMSE:** root-mean-square error
**SPCS:** State Plane Coordinate System
**TBM:** temporary benchmark
**TM:** transverse Mercator
**TS:** tangent to spiral
**USGS:** U.S. Geodetic Survey
**UTM:** Universal Transverse Mercator
**WGS84:** World Geodetic System of 1984

# INDEX

Accuracy standards, ASPRS: as USACE standard 8; classes of 9; limiting RMS error in 10; planimetric feature coordinate accuracy 10; vertical (elevation) accuracy 10–11. *See also* Accuracy standards, mapping

Accuracy standards, mapping: introduction to 3; ASPRS standards. *See* Accuracy standards, ASPRS; contour interval 3, 9; horizontal accuracy (USACE) 11; industry standards, list of 3, 8; intended use, criteria for 8; planimetric feature coordinate accuracy 10; project area 8; scale (target, map) 3, 8–9; surveying standards, independence of 3; USACE standards. *See* Mapping/surveying requirements, USACE; vertical (elevation) accuracy (ASPRS) 10–11. *See also* Accuracy standards, surveying

Accuracy standards, surveying: introduction to 12; closed-loop traverse, checking 18; control points, use of 18–19; geodetic azimuth 16; GPS coordinate transformations in 16, 18; GPS survey control 14; GPS vertical component transformations 18; horizontal control (point closure standards) 12–13; horizontal curvature correction 15; National Geographic Reference System (NGRS) 12; plane coordinate systems 14–15; primary survey control (geodetic, SPCS) 14; project datum example 15, 16; reconnoitering/planning phase 13; scale factor considerations 15–16; secondary control traverses 14; USACE standards. *See* Mapping/surveying requirements, USACE; vertical control (point closure standards) 13. *See also* Accuracy standards, mapping

Archeological site mapping (USACE) 6

As-built surveys 21–22

ASPRS standards. *See* Accuracy standards, ASPRS

Atmospheric corrections 36

Battery maintenance/power requirements 32–33

CADD data input/plotting 57. *See also* Data collector-surveyor interface

Circle eccentricity 34–35

Circle graduation error 35

Civil works GIS feature mapping (USACE) 6

Closed loop traverses: as secondary control traverses 14; example of 16

Closed traverses: as secondary control traverses 14; example of 16; point closure standards (horizontal control) 12–13; point closure standards (vertical control) 13

Coding, data: coding field data 37–38, 50; coordinate file coding scheme, USACE 54–57

Collimation error, horizontal 35

Computers, field: modems for 39; performance requirements for 38; software criteria for 38. *See also* Data collector-surveyor interface; Map compilation

Contouring: contour interval 3, 9; contour lines, drawing/weighting 27–28; contour skeleton outline 26–27; from a traversed line 26; from station 26; in open country 26

Contracts, architect-engineer: introduction to 72; alternative clauses/provisions/options 73; Contract Schedule, typical 74–85; format requirements for 73; IDT contracts 73; preparation of 65; scope of work 65; technical specifications, insertion of 72–73

Coordinate Geometry (COGO) software: introduction to 88; area computation commands 89; curve commands 89; forward computation commands 88; intersection commands 89; inverse computation commands 89; program requirements 88; typical software packages 88

Curvature: COGO software curve commands 89; horizontal curvature correction 15. *See also* Route surveying

CVTPC data conversion software 58–59. *See also* Map compilation

Data collection procedures, total station: introduction to 41; coding field data 37–38, 50; collection techniques for planimetric features 45–46; data collection sequence 43–44, 45; data flow, field-to-finish 31; detail sketch, typical 45; field book sketches, typical (digital survey) 42, 43; field crew responsibilities 46–48; field notes for data collector, typical 46; field notes for point locations, typical 47; field-to-finish procedures/flow chart 51, 52; generic data collector, functional requirements for 41–42, 50, 51; planimetric features, sketching 44–45; radial survey, establishing controls for 43; radial surveys, performing 43. *See also* Data collector-surveyor interface

Data collector-surveyor interface: CADD interface/data input 57; CADD plotting 57; coding field data 37–38, 50; computer interfacing, post-processing for 48; coordinate file coding scheme, USACE 54–57; data translation software, testing 50; digital data 48; digital data transfer 48–50; digital deliverables, anticipating 49; digital media, selection of 50; file compatibility, problems with 48; generic data collector, functional requirements for 41–42, 50, 51; graphics data, direct translation of 48–49; graphics data, neutral translators for 49; hardware compatibility 48; software selection/maintenance 49. *See also* Map compilation

Data collectors: Geodimeter 126/400 Series 51–52; Lietz SDR Series 52–53; Topcon FC-4 53; Wild GRE and REC Series 53

Data flow: field-to-finish (total station) 31; topographic, overview of 59, 60

Degree of curve (in route surveying): arc definition 67; chord definition 67; curve formulas 67–68. *See also* Curvature; Route surveying

Details, locating/plotting (plane-table surveys) 27, 29–30

Electronic theodolites, error sources for 34–35

Elevation (vertical) accuracy (ASPRS) 10–11

Error sources: combines error sources (500-ft. line) 40; controlling (checklist for) 36–37; elevation error vs. zenith angle error 39; for electronic theodolites 34–35; in total

station system 35–36; in trigonometric leveling 39; limiting RMS error (ASPRS) 10
Flood control/floodplain mapping requirements (USACE) 6
Geotechnical/hydrographic site investigation (USACE) 5
GPS (Global Positioning System): as NAD 83-referenced system 14; control points, use of 18–19; GPS survey control 14; vertical components transformation 18; WGS 84 coordinate transformation in 16, 18
Guide specifications, contract. See Contracts, architect-engineer
Hazardous/toxic waste site mapping requirements (USACE) 7
Heating, instrument (as error source) 36
Height of standards error 35
Map compilation: ASCII file, typical 59; civil/site levels (USACE) 62; CVTPC data conversion software 58–59; element symbology (USACE) 63; feature level assignments 61; field shot information, typical 59; file descriptions 59; point descriptors, typical 59, 61–62, 64; topographic data flow, overview of 59, 60
Mapping standards. See Accuracy standards, mapping; Mapping/surveying requirements, USACE
Mapping/surveying requirements, USACE: archeological site mapping 6; civil works GIS feature mapping 6; coordinate file coding scheme, USACE 54–57; emergency operation management activities 7; flood control/floodplain mapping 6; geotechnical/hydrographic site investigation 5; hazardous/toxic waste sites 7; horizontal accuracy check 11; military construction 4–5; real estate activities 6; river/harbor navigation projects 5
Metric conversions 1
Military construction mapping requirements (USACE) 4–5
Modems 39
NAD 27 (North American Datum 1927): as GPS reference system 14; as sea level reference datum 15–16
NAD 83 (North American Datum 1983): as GPS reference system 14; metric units for 1
NAVD 83 (North American Vertical Datum 1983): as vertical reference datum 14
Nets (secondary control traverses) 14
NGRS (National Geographic Reference System): in topographic survey control 12
Optical plummets/tribrachs error sources 36
Plane-table surveys: introduction to 22; details, locating/plotting 27, 29–30; equipment checklist 28; notekeeping 29; plane-table topography 22; resection in 23–24; setup hints 28; stadia traverse in 25; three-point orientation 25; traverse in 25; triangulation in 22–23; two-point problem, solving 24. See also Contouring; State Plane Coordinate System (SPCS)
Planimetric feature coordinate accuracy 10
Point closure standards: for horizontal control 12–13; for vertical control 13
Pointing errors 36

Productivity comparison: surveying methods compared 34, 35
Real estate activities, mapping requirements for (USACE) 6
Records and computations: in plane-table surveys 23
River/harbor navigation projects mapping requirements (USACE) 5
Route surveying: central angle/deflection relationship 67–68; curve formulas, required parameters for 67; deflection angles 66–67; degree of curve - arc definition 67; degree of curve - chord definition 67; horizontal circular curves 66; spiral stakeout 68–69; transition spirals 68; vertical curve geometry 69; vertical curve obstructions 70; vertical curves, equation method for 70; vertical curves, tangent offset method for 69–70
Scale: map 8–9; target 3
Sea level, reduction to (NAD 27 reference) 15–16
Site plan surveys 20–21
Specifications, contract. See Contracts, architect-engineer
Standards. See Accuracy standards (various)
State Plane Coordinate System (SPCS): in primary survey control 14; scale factors in 15–16; worksheet for 17
Station elevation, determining (plane-table surveys) 23
Surveying standards/control. See Accuracy standards, surveying
Topographic features, contouring (plane-table surveys) 23
Total station survey system: advantages of 30–31; battery maintenance/power requirements 32–33; coding field data in 37–38, 50; data flow, field-to-finish 31; electronic theodolite error sources 34–35; equipment maintenance 32; error sources, avoiding 35–36; field computers in 38; field equipment inventory 31–32; field-to-finish procedures/flow chart 51, 52; job planning/estimating 33–34; multiple rodmen, advantages of 34; productivity comparison (field time saved) 34, 35; system error control (and checklist) 36–37. See also Data collection procedures, total station
Transverse Mercator (TM) System 14
Triangulation, plane-table 22–23. See also Plane-table surveys
Trigonometric leveling: advantages/accuracies of 38–39; error sources in 39; procedures for 39
U.S. Survey Foot, metric conversion for 1
Universal Transverse Mercator (UTM) System 14–15
USACE standards. See Mapping/surveying requirements, USACE
Utility surveys 21
Vertical (elevation) accuracy: vs. zenith angle error 39; ASPRS standard for 10–11; GPS vertical component transformation 18; NAVD 83 as vertical reference datum 14; point closure standards for 13; vertical angles (in plane-table triangulation) 23; vertical curves. See Route surveying
Vibrations as error source 36
WGS 84 (World Geodetic System 1984): GPS coordinate transformation in 16, 18
Zenith angle error: vs. elevation error 39